T0133711

A Concise
Introduction to
Statistical
Inference

A Concise Introduction to Statistical Inference

Jacco Thijssen

The University of York, UK

CRC Press
Taylor & Francis Group
Boca Raton London New York

CRC Press is an imprint of the
Taylor & Francis Group, an **informa** business

A CHAPMAN & HALL BOOK

CRC Press
Taylor & Francis Group
6000 Broken Sound Parkway NW, Suite 300
Boca Raton, FL 33487-2742

© 2017 by Taylor & Francis Group, LLC
CRC Press is an imprint of Taylor & Francis Group, an Informa business

No claim to original U.S. Government works

Printed on acid-free paper
Version Date: 20171201

International Standard Book Number-13: 978-1-4987-5596-2 (Hardback)

This book contains information obtained from authentic and highly regarded sources. Reasonable efforts have been made to publish reliable data and information, but the author and publisher cannot assume responsibility for the validity of all materials or the consequences of their use. The authors and publishers have attempted to trace the copyright holders of all material reproduced in this publication and apologize to copyright holders if permission to publish in this form has not been obtained. If any copyright material has not been acknowledged please write and let us know so we may rectify in any future reprint.

Except as permitted under U.S. Copyright Law, no part of this book may be reprinted, reproduced, transmitted, or utilized in any form by any electronic, mechanical, or other means, now known or hereafter invented, including photocopying, microfilming, and recording, or in any information storage or retrieval system, without written permission from the publishers.

For permission to photocopy or use material electronically from this work, please access www.copyright.com (http://www.copyright.com/) or contact the Copyright Clearance Center, Inc. (CCC), 222 Rosewood Drive, Danvers, MA 01923, 978-750-8400. CCC is a not-for-profit organization that provides licenses and registration for a variety of users. For organizations that have been granted a photocopy license by the CCC, a separate system of payment has been arranged.

Trademark Notice: Product or corporate names may be trademarks or registered trademarks, and are used only for identification and explanation without intent to infringe.

Visit the Taylor & Francis Web site at
http://www.taylorandfrancis.com

and the CRC Press Web site at
http://www.crcpress.com

Printed and bound in the United States of America by
Edwards Brothers Malloy on sustainably sourced paper

To my family

Contents

List of Figures

Preface

From cases of miscarriage of justice to the 2008 financial crisis, the consequences of an incorrect use of statistics can be grave. Statistical inference is used as a tool to analyse evidence in the face of uncertainty. Often the mistakes in its use that lead to bad outcomes are not mathematical mistakes, in the sense that someone made a calculation error somewhere along the line. Rather, the way a problem is framed and interpreted is often at fault. When using statistics you should not just focus on the mechanics of statistical computations, but also on the setting within which these statistical tools are used. This book prepares you, not just for the nitty-gritty of statistical computation, but also for developing the ability to judge whether the appropriate computations have been made.

This book is, therefore, not only for students who aim or expect to be producers of statistical analyses, but also for those who expect to become consumers of statistical analyses. For the latter group it is very important that you have some idea about how the reports that you will read have been constructed. It's the same with driving a car: you may not be a car mechanic, but in order to refresh the oil you must have some idea about what it looks like under the hood.

The reader I have in mind is either a mathematically relatively sophisticated first-time student of statistics, or someone who has seen some statistics in the past, but not at the level of abstraction that (s)he now requires. For the former group, there are not many mathematically rigorous books that devote much attention to the modeling aspects of statistics. For the latter group, even though most advanced econometrics texts have a statistical appendix, this usually does not give a coherent enough picture as to how to interpret statistical evidence. Both groups of readers will benefit from this book, because I have kept the material, while mathematically fairly rigorous, as free from calculus as possible. The notation in the book is abstract, but most manipulations don't require more than high school algebra. Readers who are competent users of calculus can delve deeper into the theory through some of the exercises.

The main goal of the book is to help you become a competent *interpreter of evidence in the face of uncertainty*. In writing it, I have had four guiding principles in mind, which I want to share with you to make sure you can use the book to your best advantage.

An emphasis on concepts

I emphasize the *rhetoric*[1] of statistical inference right from the start. For example, rather than presenting probability and distribution theory in isolation, I introduce them entirely in the context of statistics. Also, since many students often find statistical ideas baffling and find it difficult to separate statistical ideas from the mechanics of computations, I put an emphasis on the main statistical concepts (confidence interval, hypothesis test, p-value) right from the start. Once you understand the basic ideas, the mathematics follows fairly easily. I do this by introducing the main statistical ideas early in the book (in Chapter 4) in the context of one specific statistical model. After discussing the main ideas rather informally, I then proceed by making them precise and more broadly applicable in later chapters.

In a way this is, therefore, an old-fashioned book. I do not use any computer software anywhere and all you need to do the exercises and problems are an old-fashioned calculator and some statistical tables.[2] Of course, in order to become a proficient statistician you need to learn statistical software like SPSS, Stata, or R. The danger of focussing on computer implementation too soon is that you lose sight of the foundations and, ultimately, the goal of your particular statistical analysis. The 2008 financial crisis is a good example: some of the financial products that were sold were so complex that many practitioners could no longer see the basic modeling errors that underpinned many of them.

Statistics as a toolbox to analyse uncertainty

A common misconception about statistics is that it gives some sort of certainty, because numbers don't lie. It is important to realize that statistics does not make claims as to what is "the truth." Statistics uses the mathematics of probability theory to formulate ideas and concepts that allow us to interpret evidence in the face of uncertainty. The uncertainty does not disappear because of the use of statistics. Rather, it is put in a framework that allows us to deal with it in a consistent way. In fact, there isn't just *one* framework. The one that gets most attention in this book is known as *frequentist statistics*, simply because it is most often used in the social sciences. However, in order to understand these statistical techniques more deeply, I also introduce you to another school of thought: Bayesian statistics. Bayesian statisticians use the same rules of probability theory as frequentists, but in a different way. Many students actually find Bayesian ideas more intuitive than frequentist ideas.

---

[1] In the meaning of "the art of effective or persuasive speaking or writing" (Concise Oxford dictionary).

[2] Take a look, for example, at Problems 2.7 and 2.8, where you can analyse some of the aforementioned cases of miscarriage of justice. Or Problem 2.9, where you take a look at what happens when complex financial instruments are not modeled correctly.

The role of modeling and reporting

In contrast to many other statistics texts, statistical *modeling* of real-world phenomena and *reporting* of resulting statistical analyses play important roles in the book. Statistical analyses do not exist in a vacuum. In order to get useful inferences, the statistical framework in which you study a real-world phenomenon has to be appropriate. Since any model of the real world is an abstraction that leaves out many important aspects, there are usually several models that you could conceivably use. Being able to formulate and judge different models is of the utmost importance. Therefore, I use many models that are not commonly encountered in typical statistics texts for economics, business, and the social sciences.

Once a statistical analysis has been conducted, its results have to be communicated to its users. This too requires training and I provide you with ample opportunity to do so. If, after studying this book, you will always ask yourself "what model did they use?" whenever you are confronted with a statistical analysis, then I have achieved my goal.

The book's length

The book is fairly short and only focusses on a few basic statistical techniques. It is much more important to study a few models in depth than to study a cooking book's worth of recipes that you can't fully fathom in an introductory course. This focus should also help you to come to grips with the mathematics. The concepts used in statistics are actually not that difficult; it's the mathematical treatment that students often find difficult. The mathematical treatment is necessary, though, to make statistics a truly transferrable skill. Many students ask me: "I found some [non-mathematical] book [with lots of pictures] which is much more accessible; can I use that one instead?" The answer is: no. There are many books that are more accessible, like there are some newspapers that are more accessible than others. Just because something is more accessible does not mean that it is better in achieving a goal. And the goal here is to introduce you to the statistical toolbox in such a way that you can use it flexibly. It's like learning a foreign language: rote learning a few phrases may get you through your holidays, but it's not enough to conduct a business meeting.

How to use this book

Statistics is not a spectator sport; it can only be successfully mastered if you apply it. With my students, I like to make an analogy about gym membership. Gym instructors show you how the machines work so that you can use them to improve your fitness without getting injuries. If you then choose to never use them but sit at the bar eating crisps and drinking beer instead, you can't blame the instructor for not getting fit. No pain, no gain.

Therefore, the most important parts of the book are the exercises and problems. There is no way that you are going to master statistics without trying to apply it yourself. That means that you have to invest a sufficient amount of time in trying to solve exercises and problems on your own, without looking at pre-cooked solutions. It's like learning to play the piano: watching a virtuoso play does not make *you* a good pianist. That requires effort on your side.

Before you can get to work yourself, you have to learn some basic theory, which is done in the main body of the book. My approach is to view a typical statistical analysis as consisting of three components:

1. turn the problem to be studied into an appropriate statistical model;

2. analyse the model using the available data;

3. discuss the results of the analysis and put them in context for the users of your analysis.

In analysing real-world problems you need to pay attention to all three. I call this the **MAD** (model-analyse-discuss) procedure.

Since a successful statistical analysis requires a good understanding of statistical techniques, you need

1. a basic understanding of the (mathematically formulated) statistical concepts, and

2. an ability to perform mathematical derivations and calculations.

The first step refers to the process of turning concepts into a mathematical abstraction. The way to study this aspect is to ask yourself each time you see a definition or a result/theorem: what does this mathematical statement actually say in words? This is not an easy skill to acquire, but it pays off to work on it. If you know what a mathematical statement says, then you do not have to memorize it anymore. After all, you *understand* what the statement says and, therefore, you can *reconstruct* (rather than regurgitate) it. Again, it's like learning a language: if you know what the direct object of a sentence is at an abstract level, you can, after some practising, find it in any sentence I put in front of you, rather than only in a list of memorized examples.

In the second step we use the rules of probability theory to derive results that are useful in manipulating the concepts that we have defined in the first step. In this book, we are not using anything that you have not seen before.[3] Still, many students find the manipulations difficult. I suspect that is because they have not spent enough time on the first step. After all, how can you follow the *how* if you don't know the *what*? In the language learning analogy: this second step is the actual construction of sentences in a foreign language based on your knowledge of grammar and vocabulary.

---

[3] Some parts require calculus but these parts are clearly indicated and can be omitted if so wished.

The MAD procedure allows you to apply the theory that we develop to "real-world" problems. The problems and applications that you will see in this book are highly stylized and perhaps not always very realistic. That is simply because in order to analyse *real* real-world problems we need more than just the basic theory presented here. In order to help you with the task of mastering statistics, I try to construct exercises in such a way that all three parts of the MAD learning process are clearly indicated. The problems are divided into *exercises* and *problems*. **Exercises** ask you to (i) perform basic computational tasks, (ii) prove or verify results that are mentioned in the text, or (iii) develop extensions to the theory presented. **Problems** allow you to *use* the theory that you have learned in stylized situations that are similar to problems that you may be confronted with in your career.

Exercises that are marked (∗) are usually more challenging, of a more theoretical nature, or use differential or integral calculus and should appeal to students with an interest in the theoretical development of statistical methodology. If you are more interested in *using* rather than *developing* statistical tools, you can omit these exercises. If, on the other hand, you like a mathematical challenge, then these exercises allow you to extend your statistical toolbox a bit beyond the standard material.

Throughout the text, I use symbols in the margin to help your learning.

In a section marked with a coffee cup, I provide some additional thoughts, background, intuition, or some computations that require more thought and effort. You may want to get yourself a cup of coffee or tea before studying these sections.

When you encounter the "dangerous bend" sign, you should take extra care, because what follows is quite tricky, or describes a source of many mistakes.

A section marked with the integral symbol indicates that the paragraph $\int$ that follows uses differential and/or integral calculus. If you are not conversant with calculus, you can read around the calculations; the concepts and flow of the theory should still be accessible.

Acknowledgments

Many people have helped in writing this book. First of all, my own teachers, in particular Feico Drost, Robert de Jong, Bertrand Melenberg, and Harry Tigelaar. Throughout my teaching statistics I have been fortunate to have been working with very helpful, talented, and supportive teaching assistants: Augustin Benetrix, Rory Broderick, Karol Borowiecki, John Frain, Emma Howard, Julia Anna Matz, Denis Tkachenko and Stefano Verde at Trinity College Dublin, and Daniele Bregantini and Yaprak Tavman at the University of York.

Many colleagues have helped in the development of the book through direct comments and/or discussions about statistics: in particular, Gustav Delius, Fabrizio Iaccone, Peter Lee, and Brendan McCabe. I would also like to express my gratitude to Simon Eveson for his assistance with producing many of the graphs using the wonderful LaTeX package TikZ.

At Taylor & Francis I would like to thank Rob Calver for giving me the opportunity to embark on this project. Rebecca Davies, Alex Edwards, and Robin Lloyd-Starkes are thanked for their guidance and support throughout the production process, and especially for their patience with my difficulties in letting go of the book. "Let me just make one more change!" Marcus Fontaine's assistance with LaTeX issues is gratefully acknowledged.

Finally, and most importantly, I wish to thank my family, Ruth and Peter in particular, for their love and understanding. Peter's smile is the best reward one can get after a day of writing.

St. Senoch and York, July 2016

Chapter 1

Statistical Inference

1.1 What statistical inference is all about

The decisions of politicians, businesses, engineers, not-for-profit organizations, etc. typically have an influence on many people. Changes to child benefits by a government, for example, influence the financial position of many households. The government is interested (hopefully) in the effect of such measures on individual households. Of course, the government can't investigate the effect on every individual household. That would simply take too much time and make it almost impossible to design general policies. Therefore, the government could restrict itself by focussing on the *average* effect on households in, say, the lowest income quartile. Even finding out this number is typically too difficult to do exactly. So, the government relies on information obtained from a small subset of these households. From this subset the government will then try to *infer* the effect on the entire population.

The above example gives, in a nutshell, the goal of statistics. Statistics is *the study of collecting and describing data and drawing inferences from these data.* Politicians worry about the impact of budgetary measures on the *average* citizen, a marketeer is concerned with *median* sales over the year, an economist worries about the *variation* in employment figures over a 5-year period, a social worker is concerned about the *correlation* between criminality and drug use, etc. Where do all these professionals get that information from? Usually from data about their object/subject of interest. However, a long list of numbers does not really help these professionals in analysing their subject and in making appropriate decisions accordingly. Therefore, the "raw data" (the list of responses you get if, for example, you survey 500 people) are condensed into manageable figures, tables, and numerical measures. How to *construct* these is the aim of descriptive statistics. How to *use* them as evidence to be fed into the decision making process is the aim of inferential statistics and the subject of this book. This chapter introduces in an informal way some of the statistical jargon that you will encounter throughout the book.

Inferential statistics is the art and science of interpreting evidence in the face of uncertainty.

1

Example 1.1. Suppose that you want to know the average income of all university students in the country (for example, to develop a financial product for students). Then, obviously, you could simply go around the country and ask every student after their income. This would, however, be a very difficult thing to do. First of all, it would be extremely costly. Secondly, you may miss a few students who are not in the country at present. Thirdly, you have to make sure you don't count anyone twice.

Alternatively, you could only collect data on a subgroup of students and compute their average income as an approximation of the true average income of all students in the country. But now you have to be careful. Because you do not observe all incomes, the average that you compute is an *estimate*. You will need to have some idea about the accuracy of your estimate. This is where inferential statistics comes in.   ◁

Let's rephrase the above example in more general terms: you wish to obtain information about a summary (called a **parameter**) of a measurement of a characteristic (called a **variable**) of a certain group of people/objects/procedures/... (called a **population**) based on observations from only a subset of the population (called a **sample**), taking into account the distortions that occur by using a sample rather than the population. All of these boldface notions will be made precise in this book. For now it suffices to have an intuitive idea.

The goal of inferential statistics is to develop methods that we can use to infer properties of a population based on sample information.

1.2   Why statistical inference is difficult

There is a great need for methods to gather data and draw appropriate conclusions from the evidence that they provide. The costs of making erroneous decisions can be very high indeed. Often people make judgements based on *anecdotal evidence*. That is, we tend to look at one or two cases and then juxtapose these experiences onto our world view. But

anecdotal evidence is not evidence.

At its most extreme, an inference based on anecdotal evidence would be to play the lottery because you heard that a friend of a friend's grandmother once won it. A collection of anecdotes never forms an appropriate basis from which general conclusions can be drawn.

Example 1.2 (Gardner, 2008). On November 6, 2006, the *Globe and Mail* ran a story about a little girl, who, when she was 22 months old, developed an aggressive form of cancer. The story recounted her and her parents' protracted battle against the disease. She died when she was 3 years old. The article came complete with photographs of the toddler showing her patchy hair due to chemotherapy. The paper used this case as the start for a series of articles about cancer and made the little girl, effectively, the face of cancer. ◁

No matter how dreadful this story may be from a human perspective, it is not a good basis for designing a national health policy. The girl's disease is extremely rare: she was a one-in-a-million case. Cancer is predominantly a disease of the elderly. Of course you could say: "any child dying of cancer is one too many," but since we only have finite resources, how many grandparents should not be treated to fund treatment for one little girl? The only way to try and come up with a semblance of an answer to such questions is to *quantify* the effects of our policies. But in order to do that we need to have some idea about effectiveness of treatment in the *population as a whole*, not just one isolated case.

The human tendency to create a narrative based on anecdotal evidence is very well documented and hard-wired into our brains.[1] Our *intuition* drives us to make inferences from anecdotal evidence. That does not make those inferences any good. In fact, a case can be made that societies waste colossal amounts of money because of policies that are based on anecdotal evidence, propelled to the political stage via mob rule or media frenzy.

In order to control for this tendency, we need to *override* our intuition and use a formal framework to draw inferences from data. The framework that has been developed over the past century or so to do this is the subject of this book. The concepts, tools, and techniques that you will encounter are distilled from the efforts of many scientists, mathematicians, and statisticians over decades. It represents the core of what is generally considered to be the consensus view of how to deal with evidence in the face of uncertainty.

1.3 What kind of parameters are we interested in?

As stated above, statistics starts with the idea that you want to say something about a parameter, based on information pertaining to only a subgroup of the entire population. Keep in mind the example of average income (parameter) of all university students (population). Of course not every student has the same income (which is the variable that is measured). Instead there is a spread of income levels over the population. We call such a spread a **distribution**. The

---

[1]See Gardner (2008) or Kahneman (2011) for reader-friendly accounts of psychological biases in decision making under uncertainty.

distribution tells you, for example, what percentage of students has an income between \$5,000 and \$6,000 per year. Or what percentage of students has an income above or below \$7,000. The parameter of interest in a statistical study is usually a particular feature of this distribution. For example, if you want to have some idea about the center of the distribution, you may want to focus on the **mean** (or average) of the distribution. Because the mean is so often used as a parameter, we give it a specific symbol, typically the Greek[2] letter μ.

1.3.1   Center of a distribution

The mean, or average, of a population gives an idea about the *center* of the distribution. It is the most commonly used summary of a population. The average is often interpreted as describing the "typical" case. However, if you collapse an entire population into just one number, there is always the risk that you get results that are distorted. The first question that should be answered in any statistical analysis is: "Is the parameter I use appropriate for my purpose?"

In this book I don't have much to say about this: we often deal with certain parameters simply because the theory is best developed for them. A few quick examples, though, should convince you that the question of which parameter to study is not always easy to answer.[3] Imagine that you are sitting in your local bar with eight friends and suppose that each of you earns \$40,000 per year. The average income of the group is thus \$40,000. Now suppose that the local millionaire walks in who has an income of \$1,500,000 per year. The average income of your group now is \$186,000. I'm sure you'll agree that average income in this case is not an accurate summary of the population.

This point illustrates that the mean is highly sensitive to **outliers**: extreme observations, either large or small. In the income case it might be better to look at the **median**. This is the income level such that half the population earns more and half the population earns less. In the bar example, no matter whether the local millionaire is present, the median income is \$40,000. The difference between mean and median can be subtle and lead to very different interpretations of, say, the consequences of policy. For example, during the George W. Bush administration, it was at one point claimed that new tax cuts meant that 92 million Americans would, on average, get a tax reduction of over \$1,000. Technically, this statement was correct: over 92 million Americans received a tax cut and the average value was \$1,083. However, the median tax

---

[2]It is common in mathematics to use Greek letters for parameters. The reason is that we often have to assign symbols to many different variables, parameters, etc. We would run out of symbols very quickly if we could only use the Latin alphabet. See Appendix D for the Greek alphabet.

[3]These examples are taken from Wheelan (2013), which comes as a highly recommended account on the uses and abuses of statistics.

cut was under \$100. In other words, a lot of people got a small tax cut, whereas a few Americans got a very large tax cut.

Not that the median is always a good measure to describe the "typical" case in a population either. For example, if a doctor tells you after recovery from a life-saving operation that the median life expectancy is 6 months, you may not be very pleased. If, however, you knew that the average life expectancy is 15 years, the picture looks a lot better. What is happening here is that a lot of patients (50%) die very soon (within 6 months) after the operation. Those who survive the first 6 months can look forward to much longer lives.

Both examples arise in cases of *skewed* distributions, i.e., distributions that are not **symmetric** around the mean. By a symmetric distribution we mean a distribution where, in the example of student income, the percentage of students who exceed a certain income in excess of the mean is the same as the percentage of students who fall below the same level of income below the mean. An illustration of a symmetric and a non-symmetric distribution is given in Figure 1.1. The distribution on the right is called **skewed to the right**, because it describes a distribution where a few students earn a lot more than the mean.

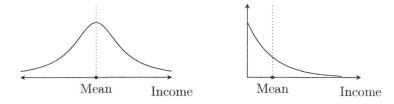

FIGURE 1.1: A symmetric and a skewed distribution.

Question 1.1. In the distribution on the right, is the mean larger or smaller than the median?

1.3.2 Spread of a distribution

Even if we are only interested in the mean of a distribution, it turns out that we also need to know something about the *dispersion* of the distribution around the mean. Are all subjects clustered together around the mean or is there a big spread? The most common way to measure the spread of a distribution is by the **variance**. This quantity is obtained by computing the average *squared* distance to the mean. By squaring the distances, we penalize large deviations more than small deviations. In Figure 1.2, you see two symmetric distributions with the same mean, but different variances.

Note that, since we take squares to obtain the variance, it does not have the same unit of measurement as the original variable. Therefore, we often take the square root of the variance to obtain what is called the **standard**

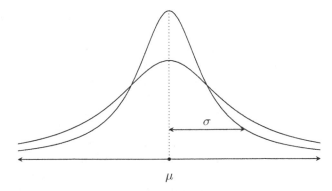

FIGURE 1.2: Two distributions with different variances.

deviation, which is usually denoted by the Greek letter σ. So, if the standard deviation is σ, then the variance is σ^2.

Even though the variance is often used as a tool when we actually want to say something about the mean (as we will do repeatedly in later chapters), the variance can be a parameter of interest in its own right. Suppose, for example, that you are the manager of a firm that manufactures jet engine fan blades. Because of its use, your product has to be produced to very stringent technical specifications. For quality control purposes you may then be interested in the variance of a particular measurement pertaining to your product.

1.3.3   Association between variables

Often we are interested in how two (or more) variables are associated. For example, the performance of university students and parental income may be associated. Getting some insight into such a relationship is important for policy makers who want to improve the performance of students. If it turns out that attainment is linked to parental income, then perhaps an early childhood intervention program may be desirable.

There are many ways in which the association between two variables can be measured. One popular way is closely related to the variance. There we computed the average of squared deviations from the mean. With two variables we could proceed by computing for each member of the population the distance to the average for each of the two variables and then multiplying. So, if a student both performs much better than average and her parents have a much higher than average income, we will be multiplying two large positive numbers, giving a large positive number. If a student performs much better than average and her parents have a slightly lower than average income, we will be multiplying a large positive and a small negative number, resulting in a negative number. If we now take the average over all these numbers, we get the so-called **covariance**. A high positive covariance indicates that large deviations from the mean in one variable are, on average, accompanied by large positive deviations from the mean in the other variable. A small negative covariance, on the other hand, indicates that deviations from the mean in the first variable are, on average, slightly more than offset by deviations in the second variable.

As with the variance, the unit of measurement of the covariance is not related to the units of measurement in either variable. To correct for this, we often divide the covariance by the product of the two standard deviations to get the **correlation coefficient**, which is often denoted by ρ. This parameter is a number between -1 and 1 and measures the strength of a linear relationship between the two variables. To visualize, imagine that, for each student, you put a cross in the plane that is obtained when you measure attainment on the horizontal and parental income on the vertical axis. (Such a plot is called a **scatter plot**.) A large positive (negative) correlation indicates that all points lie clustered fairly tightly around a straight line with positive (negative) slope. A small positive (negative) correlation indicates that all points lie clustered around a straight line with positive (negative) slope in a much more dispersed fashion.

In Figure 1.3, two scatter plots are drawn for six pairs of observations of variables x and y. In both cases the correlation is positive, because the best-fitting straight line has a positive slope. The correlation in the left panel, however, is higher than in the right panel, because the observations in the former are more closely clustered around the best-fitting straight line than in the latter.[4]

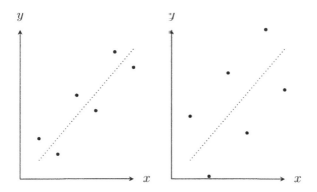

FIGURE 1.3: Two scatter plots.

---

[4]What we mean by "best-fitting straight line" will be made clear in Chapter 9.

1.4   Statistics and probability

Suppose that you want to know the average amount of caffeine in cups of coffee that you buy at your favorite coffee purveyor. You want to base your conclusions on nine cups of coffee that you will buy and analyse carefully. To put this simple example into statistical jargon, call

- population: all (potential) coffees (that could ever conceivably be) sold by the coffee shop;

- variable: the caffeine content of a particular cup of coffee;

- parameter: the average caffeine content of all coffees sold by the coffee shop;

- sample: the nine cups of coffee that you buy from the shop.

No two cups of coffee have *exactly* the same amount of caffeine, even though they may all be produced using the same machine. There is *randomness*, or *uncertainty*. If there wasn't, you could just buy one cup of coffee and you would know exactly how much caffeine each cup contains. Instead, the caffeine content of all cups of coffee has a particular distribution and you are interested in its mean.

A similar reasoning holds for the sample mean. Each group of nine cups of coffee is different and, therefore, each sample will have a different sample mean. This gives a new population: all possible sample means that you can observe in a sample of nine cups of coffee. These sample means also have a distribution. Statisticians often call this a **sampling distribution** to distinguish between the distribution of the variable (caffeine content in a cup of coffee) and the distribution of a statistic (the sample mean caffeine content of a sample of nine randomly chosen cups of coffee).

Think of it as two different worlds, one inhabited by cups of coffee (the "real world") and one inhabited by random samples of nine cups of coffee (the "sample world"). We are interested about the mean of the distribution in the real world, but we will conduct our inferences (at least in part) by looking at the sample world, i.e., the sampling distribution of the sample mean.

Suppose that the coffee shop claims that the average caffeine content is 50 mg. If you buy one cup of coffee and you find a caffeine content of 52 mg, does this mean that the company's claim is wrong? Not necessarily. After all, the shop says something about average caffeine content, not about the caffeine content of each individual cup of coffee. You could ask yourself, however, how *likely* it is that a randomly chosen cup of coffee has a caffeine content of more than 52 mg. Equivalently, if you buy nine cups of coffee, you would expect to find a sample mean pretty close to 50 mg if the company's claim is correct. If you find a sample mean of 52 mg, the company can, of course, still be correct and your observation could purely be down to chance. But you can again ask yourself how likely it is that you observe a sample mean of 52 mg in a sample of nine taken from a population of cups of coffee with an average caffeine content of 50 mg. Intuitively it should be clear that, if the company is correct, it should be less likely to find an average of at least 52 mg in a sample of nine cups than it is if you just have one cup.

But how much less likely? In order to answer that question we need to make precise what we mean by "likely." We also need to develop a theory that allows us to perform computations of the kind indicated above. That (mathematical) theory is called **probability theory** and will be the topic of the next chapter. After that, in Chapter 3, we will return to the questions posed above and use probability theory to provide some answers.

1.5 Chapter summary

Statistics is concerned with using sample information to draw inferences about population parameters. In order to deal with uncertainty, we will need to develop a theory of probability.

Chapter 2

Theory and Calculus of Probability

The theory of statistical inference is based on probability theory, the starting point of which is a probability model. This is a mathematical description of some *experiment* that is subject to randomness, such as flipping a coin, rolling a die, asking a voter which party (s)he will vote for, testing a new wing design for aircraft in a wind tunnel to measure drag, etc. This chapter introduces the parts of probability theory that are useful for (basic) statistical inference.

2.1 Probability models

When thinking about how to write down an abstract model of experiments that involve random outcomes, it helps to keep a simple example in mind. An easy one is rolling a (six-sided) die. There are three fundamental ingredients in describing this experiment. First, we need to be able to write down every outcome that can occur (here: $\{1, 2, 3, 4, 5, 6\}$). Secondly, we must be able to describe all the configurations of outcomes (which we will call "events") to which we want to assign a probability. For example, the event "an even number comes up" is described by $\{2, 4, 6\}$. Finally, we need to assign probabilities to events. For example, if the die is fair, you would assign the probability $1/2$ to the event that an even number comes up.

Abstracting from this particular example, a **probability model** is a triple $(\Omega, \mathscr{F}, \mathsf{P})$, where Ω is the set of all possible **outcomes**, and $\mathscr{F}$ is the set of all possible **events**. The set of events should satisfy some properties that ensure that we can assign probabilities to all events we might conceivably want to assign them to.

(E1) The set of outcomes is an event. In mathematical notation:[1] $\Omega \in \mathscr{F}$.

(E2) If A is an event, then so is "not A." In mathematical notation: $A \in \mathscr{F} \Rightarrow A^c \in \mathscr{F}$.

(E3) If you take any (countable[2]) collection of mutually exclusive events, then

---

[1]See Appendix E for a glossary of mathematical notation.

[2]Roughly speaking, a set is countable if you can put all its elements in a row and count them. The set of natural numbers, $\mathbb{N}$, is an obvious example. Also the set of rational

11

their union is also an event. In mathematical notation: $A_1, A_2, \cdots \in \mathscr{F} \Rightarrow \cup_{k=1}^{\infty} A_k \in \mathscr{F}$.

The final ingredient of a probability model is a mapping P that assigns to each event $A \in \mathscr{F}$ a **probability** $P(A)$. The probability P is assumed to satisfy the following axioms:

(P1) the probability of any event is non-negative, i.e., $P(A) \geq 0$, for any $\mathscr{F}$;

(P2) the event "some outcome occurs" is certain, i.e., $P(\Omega) = 1$;

(P3) if you take any (countable) collection of mutually exclusive events, then the probability of any of these events occurring equals the sum of the individual probabilities, i.e., $P(\cup_{k=1}^{\infty} A_k) = \sum_{k=1}^{\infty} P(A_k)$, for any $A_1, A_2, \cdots \in \mathscr{F}$ such that $A_k \cap A_\ell = \emptyset$, whenever $k \neq \ell$.[3]

An easy way to see that these requirements make intuitive sense is to replace the word "probability" by "percentage" in (P1)–(P3) above. Note that (P2) only makes sense because of (E1) and (P3) only makes sense because of (E3). In what follows I will not often make explicit that an event A is an element of $\mathscr{F}$; it is implicit in the claim that A is an "event."

The probability of an event A is usually interpreted as the *frequency of the event occurring in a large number of repetitions of the experiment*. Note that this interpretation is just that, an *interpretation*. There is nothing in the mathematical axiomatization that forces you to do so. In fact, there is a whole school of thought that interprets probability as a *degree of personal belief*, rather than an objective frequency.[4]

Example 2.1. Suppose that you flip a three-sided die (for example, a rounded-off triangular prism), the faces of which bear the letters a, b, and c, respectively. Then the outcome space of this experiment is $\Omega = \{a, b, c\}$. The set of all possible events is the set of all possible subsets of Ω, i.e.,

$$\mathscr{F} = \{\, \emptyset, \{a\}, \{b\}, \{c\}, \{a,b\}, \{a,c\}, \{b,c\}, \{a,b,c\} \,\}.$$

Take a little time to fully understand this: $\mathscr{F}$ is a *set of sets*. In words, the event $\{b, c\}$, for example, reads "when I roll the die either b or c comes up."

Question 2.1. Verify that $\mathscr{F}$ satisfies (E1)–(E3).

If you believe that the die is fair, i.e., that each side is equally likely to

---

numbers, $\mathbb{Q}$, is countable. The set of real numbers, $\mathbb{R}$, is uncountable. You can never put, for example, all real numbers between 0 and 1 in a row and count them. Between each two numbers in such a row there are infinitely many other numbers. This is also the reason why you cannot talk about "the smallest real number larger than zero." That number does not exist.

[3] See Appendix F for an explanation of the summation notation.

[4] See Appendix H.

come up, then an appropriate choice for the probability of an event is to simply count the number of elements in the event, divided by 3, i.e.,

$$\mathsf{P}(A) = \frac{|A|}{|\Omega|} = \frac{|A|}{3}.$$

Here $|A|$ denotes the **cardinality** of the set A, i.e., the number of elements in the set A.

Question 2.2. Verify that this probability satisfies (P1)–(P3).

Suppose now that, for some reason, you know that the outcome a is twice as likely as the other two. We will construct the appropriate probability and immediately see why mathematical notation is handy. Without it you would now have to write down every probability for every single event, of which there are $2^3 = 8$. In this example that is not too onerous, but what if we had been flipping a normal die with six sides? We would have to write down $2^6 = 64$ numbers. If the experiment was to roll a six-sided die twice, the number of probabilities to write down would be (check!) $6.87 \cdot 10^{10}$. So let's try and do this in the most efficient way possible. The first thing to realise is that the probability of any event depends, crucially, on the probabilities of each side coming up. Let's give these numbers a name, say, $p_a = 1/2$, $p_b = 1/4$, and $p_c = 1/4$, respectively. I could just as well have called them "Tom," "Dick," and "Harry," but it's of course much easier to keep in mind what they represent by giving them the names I did: p for probability and the subscript indicating which side is referred to.[5] How should we now assign a probability to an event A? Take, as an example, $A = \{a, c\}$. The probability of rolling an a or a c is simply the sum of the individual probabilities (because these events are mutually exclusive), i.e., $\mathsf{P}(A) = p_a + p_c = 3/4$. What we've done is to find which sides k are in the event A and then add the corresponding p_k's. In slightly fewer words: we look at the set $\{\, k \in \Omega \mid k \in A \,\}$ and sum all these p_k's. In even fewer words: for any $A \in \mathscr{F}$, we define

$$\mathsf{P}(A) = \sum_{\{\, k \in \Omega \mid k \in A \,\}} p_k.$$

$\triangleleft$

All results and theorems in probability theory can be derived from the three basic axioms mentioned above. A few of those rules follow below.

Proposition 2.1. *Let* $(\Omega, \mathscr{F}, P)$ *be a probability model. Then the probability* P *satisfies the following properties.*

---

[5] Non-mathematically trained students often find mathematical notation baffling, but you should remember that all you're doing here is giving an object a name. So you might as well choose a name that makes some intuitive sense.

(i) *If A is an event that is contained in B (in other words, B holds whenever A holds), then the probability of event B exceeds that of event A, i.e., $P(A) \leq P(B)$, whenever $A \subseteq B$.*

(ii) *For any event A, the event "A or not A" is sure to occur, i.e., $P(A) + P(A^c) = 1$, for all $A \in \mathscr{F}$.*

(iii) *The probability of any event does not exceed 1, i.e., $P(A) \leq 1$, for any $A \in \mathscr{F}$.*

(iv) *The event "no outcome occurs" certainly does not happen, i.e., $P(\emptyset) = 0$.*

(v) *For any two events, the probability of at least one event occurring is the sum of the probabilities of the two events net of the probability of both events occurring simultaneously, i.e., $P(A \cup B) = P(A) + P(B) - P(A \cap B)$, for any $A, B \in \mathscr{F}$.*

A good way of understanding the axioms (P1)–(P3) and the results that are derived from them is by drawing a Venn diagram. You will be asked to prove these statements in Exercise 2.1, but, as an example, let's prove the second one.

Proof. Take any $A \in \mathscr{F}$. From (E2) we know that $A^c \in \mathscr{F}$, so that $P(A^c)$ exists. Note that A and A^c are mutually exclusive events, so that (P3) implies that $P(A \cup A^c) = P(A) + P(A^c)$. Since $A \cup A^c = \Omega$, (P2) then implies that $P(A) + P(A^c) = 1$. ∎

If you think about the probability of an event as corresponding to the percentage of the population with that property, then all these statements should make perfect sense.

The **conditional probability** of an event A, given an event B, is defined as

$$P(A|B) = \frac{P(A \cap B)}{P(B)}, \tag{2.1}$$

provided that $P(B) > 0$. It can be shown (Exercise 2.2) that the mapping $P(\cdot|B)$ is itself a probability (i.e., satisfies the three axioms (P1)–(P3)).

Two events A and B are called **independent** if $P(A \cap B) = P(A)P(B)$. A more intuitive way of thinking about independence can be found by rewriting this definition, which gives that two events are independent if $P(A|B) = P(A)$. In other words, if A and B are independent, then your probability assessment of A does not change if I tell you that B has occurred.

Suppose that $(A_1, \ldots, A_n)$ is a collection of mutually exclusive ($A_i \cap A_j = \emptyset$, $i \neq j$) and collectively exhaustive ($\cup_{i=1}^n A_i = \Omega$) events. The rule of **total probability** states that, for any event $B \in \mathscr{F}$, it holds that

$$P(B) = \sum_{i=1}^n P(B \cap A_i) = P(B \cap A_1) + \cdots + P(B \cap A_n).$$

Using the definition of conditional probability (2.1), we can rewrite this as

$$P(B) = \sum_{i=1}^{n} P(B|A_i)P(A_i) = P(B|A_1)P(A_1) + \cdots + P(B|A_n)P(A_n).$$

The rule of total probability also shows up in a result known as **Bayes' rule** (named after the Rev. Thomas Bayes, 1701–1761). This rule is a straightforward application of the definition of conditional probability and the rule of total probability:

$$P(A_k|B) = \frac{P(B|A_k)P(A_k)}{\sum_{i=1}^{n} P(B|A_i)P(A_i)}, \quad k \in \{1, 2, \ldots, n\}.$$

Bayes' rule has many interesting applications, some of which you will see in the Problems section. It also lies at the heart of a school of statistical thought that will be briefly introduced in Chapter 10.

2.2 Random variables and their distributions

In statistics we are usually not interested in the actual experiment, but in some *measurable outcome* of the experiment. For example, suppose that you ask 63 voters whether they agree with a certain policy initiative. What you are interested in here is not so much the actual experiment of asking 63 voters for their agreement, but the number of these 63 voters who agree. Let's denote this number by X. Note that X can take several values $(0, 1, 2, \ldots, 63)$ and that, before we actually ask the voters (i.e., "conduct the experiment"), we don't know what value X is going to take. That makes the variable X a **random variable**.

This is actually very unfortunate terminology, because a random variable is not a variable, but a *function*. After all, remember what we're doing: we are conducting an experiment $(\Omega, \mathscr{F}, P)$ and once we see the outcome write down a particular characteristic. So, the random variable X maps the outcome into a real number, i.e., $X : \Omega \to \mathbb{R}$. Every random variable is characterized (completely determined) by its **distribution function**, F. This function assigns to every real number the probability that the random variable takes an outcome at most equal to that number. So, $F_X : \mathbb{R} \to [0, 1]$ is such that, for any $x \in \mathbb{R}$,

$$F_X(x) = P(\{X \le x\}).$$

Note that F_X is a non-decreasing function. I tend to use the subscript X to make sure that I do not forget which random variable I'm looking at.

Before moving on, let's spend some time on the probability statement in the above equation. The event that we are looking at is the event "X does not exceed x." Often we are a bit sloppy, leave out the brackets and write $P(X \leq x)$. But wait a minute; the brackets indicate that we are dealing with a *set*. What set is that? Remember that the probability P is defined on events in the set of events $\mathscr{F}$. And we defined events as subsets of the outcome space Ω. So, what event is $\{X \leq x\}$? This is where it becomes important that we realize that X is a function. We are looking at all possible outcomes of the experiment and then taking only those that lead to a value of X that does not exceed x. In mathematical notation, we are looking at the set $\{\,\omega \in \Omega \mid X(\omega) \leq x\,\}$. When we write $\{X \leq x\}$ we actually use this as shorthand notation.[6]

Two random variables X and Y are **independent** if the events $\{X \leq x\}$ and $\{Y \leq y\}$ are independent for all x and y. In other words, X and Y are independent if we can write the joint probability of $\{X \leq x\}$ and $\{Y \leq y\}$ as the product of the distribution functions, F_X and F_Y, respectively:

$$P\left(\{X \leq x\} \cap \{Y \leq y\}\right) = F_X(x)F_Y(y).$$

Random variables are often classified as **discrete** or **continuous**. A discrete random variable is one that takes countably many possible values (for example, in the sets $\{0, 1\}$, $\mathbb{N}$, or $\mathbb{Z}$). A discrete random variable is not only characterised by its distribution function, but also by its **probability mass function**, which we denote by f_X. Let $\mathscr{R}$ be the range of X. Then,

$$f_X(k) = P(X = k), \quad k \in \mathscr{R}.$$

[Note that I have been somewhat sloppy again and should have written $P(\{X = k\})$ or, even better, $P(\{\omega \in \Omega | X(\omega) = k\})$.]

A continuous random variable takes value in an uncountable set (such as the interval $[a, b]$ or the real line $\mathbb{R}$). A continuous random variable[7] is characterized by its **density**, which we will denote by f_X. Let $\mathscr{R}$ be the range of X. The density is a function f_X such that

$$\int_a^b f_X(x)dx = F_X(b) - F_X(a),$$

for any $a, b \in \mathscr{R}$, where F_X is the distribution function of X. Note that I use the same symbol for mass and density functions. I do this to emphasize that they play a similar role, albeit for different types of random variables. In fact, I will often use the term "density" to mean both density and mass functions.

---

[6] For those with a mathematical interest, you may have spotted that I have been a bit sloppy. After all, there is no *guarantee* that $\{X \leq x\}$ is an event, i.e., that $\{X \leq x\} \in \mathscr{F}$. In fact, that is part of the definition of a random variable. So, a random variable is formally *defined* as a function $X : \Omega \to \mathbb{R}$, *such that* $\{X \leq x\} \in \mathscr{F}$, *for all* $x \in \mathbb{R}$. It's actually even more complicated than this, but we won't worry about that here.

[7] Technically speaking, I should call these random variables *absolutely continuous*. Throughout this book, the mathematically informed reader should read "absolutely continuous random variable" when (s)he sees "continuous random variable" written down.

Let's try and understand this definition a bit better. From the definition of the distribution function, we see that

$$F_X(b) - F_X(a) = \mathsf{P}(X \le b) - \mathsf{P}(X \le a) = \mathsf{P}(a \le X \le b).$$

So, the area under the curve of f_X between a and b is the probability that X takes a value between a and b.

In statistics we often bypass the probability model altogether and simply specify the random variable directly. In fact, there is a result (which we won't cover in detail) that tells us that, for any random variable, we can find an appropriate probability model. So, from now on we will mostly forget about the probability model, but remember that it is always there lurking in the background.

Example 2.2. Consider the example of asking n voters if they agree with a particular policy. Let X denote the number of voters who agree. Let's assume that each voter agrees with a constant probability p. So, p represents the fraction of all voters in the population who agree with the policy. Suppose that each voter is drawn independently and **with replacement** from the population of all voters. That means that we conduct our experiment in such a way that each voter in the population is equally likely to be sampled and that each sampled voter is returned to the population after sampling and can, thus, be asked again. We can now write down the probability mass function, i.e., the probability that the random variable X takes a particular value, say, k.

If k voters agree, then $n-k$ voters must disagree. The probabilities of each voter agreeing and disagreeing are p and $1-p$, respectively. Since all voters are drawn independently, we can compute the probability that k voters agree and $n-k$ voters disagree by multiplying: $p^k(1-p)^{n-k}$. But now we have to think about how many ways there are to draw k voters out of n. For example, it could be that the first k voters all agree and all the others disagree. Or maybe the first $n-k$ voters disagree and the final k agree. Or maybe the first voter agrees, then $n-k$ disagree and the final $k-1$ agree again, etc. The total number of ways of choosing k out of n is

$$\binom{n}{k} = \frac{n!}{k!(n-k)!},$$

where $n!$ (pronounced "n factorial") is $n! = n \cdot (n-1) \cdot (n-2) \cdots 1$. This so-called **binomial coefficient** is also often denoted as C_n^k or $_nC_k$. So, we can now write that

$$f_X(k) = \mathsf{P}(X = k) = \binom{n}{k} p^k (1-p)^{n-k}.$$

This completely characterizes the random variable X. ◁

In fact, we use the distribution described in the example above so often that it gets a special name: we call it the **binomial** distribution. Notice that it is completely determined by two numbers: the number of voters sampled, n, and the probability of a voter agreeing, p. We call these the *parameters* of the distribution. A shorthand notation for describing that a random variable X follows a binomial distribution with parameters n and p is

$$X \sim \mathsf{Bin}(n, p).$$

To return to the frequentist interpretation of probability, by saying that $X \sim \mathsf{Bin}(n, p)$ we essentially state that a fraction $\binom{n}{k} p^k (1-p)^{n-k}$ of random samples drawn with replacement of size n will have exactly k voters agreeing with the policy. The mass functions of three different binomial distributions are plotted in Figure 2.1.

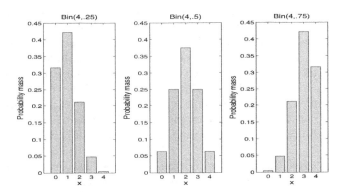

FIGURE 2.1: Probability mass function of three different binomial random variables.

So, if $X \sim \mathsf{Bin}(n, p)$, then X is the number of successes in n independent trials of some experiment. We can describe the outcomes of individual trials as well. Let X_i denote the outcome of trial i. Then $X_i \in \{0, 1\}$ and $\mathsf{P}(X_i = 1) = p$ (and, thus, $\mathsf{P}(X_i = 0) = 1 - p$). A random variable with this distribution is said to have a **Bernoulli** distribution and is denoted by $X_i \sim \mathsf{Bern}(p)$. It should be obvious now that that, if $X_1, \dots, X_n$ are independent Bernoulli random variables, then $\sum_{i=1}^{n} X_i \sim \mathsf{Bin}(n, p)$.

Example 2.3. Consider the same setting as in Example 2.2, but now assume that we can only sample each individual from the population at most once. That is, we sample **without replacement**. If the population consists of N individuals, this means that Np agree and $N(1-p)$ disagree with the policy. What is the probability that k voters out of n sampled voters agree? First note that there are $\binom{N}{n}$ ways in which we can draw a sample of size n out of a population of N voters. Since there are Np voters who agree, there are $\binom{Np}{k}$

ways in which we can draw k voters who agree and there are $\binom{N(1-p)}{n-k}$ ways in which we can draw $n - k$ voters who disagree in a sample of size n. So, if X denotes the random variable counting the number of voters in a sample of n agreeing with the policy, then its probability mass function is

$$f_X(k) = \mathsf{P}(X = k) = \frac{\binom{Np}{k}\binom{N(1-p)}{n-k}}{\binom{N}{n}}.$$

$\triangleleft$

Again, this is a distribution that is used so often that it is given a special name: the **hypergeometric** distribution. It depends on three parameters: n, N, and p, and is denoted $X \sim \mathsf{H}(n, N, p)$. It should be intuitively clear that, if your population is very large, then sampling with or without replacement should not make much of a difference, because the probability of sampling someone twice is very small. Indeed, it can be shown that, as $N \to \infty$, the binomial and hypergeometric distributions coincide.

There are many standard distributions, one of which will be studied in detail in Section 2.5. For an overview of the most common discrete and continuous distributions, see Appendices A and B, respectively.

2.3 Moments of random variables

As we saw in the previous chapter, we often want to collapse a characteristic of an entire distribution into one or two descriptive statistics that, we hope, give an accurate picture of the population. Two of the most basic characteristics of distributions are *centrality* and *dispersion*.

2.3.1 Expectation

As a measure for the centre of a distribution, we can use the mean, or average, of a distribution (taking into account the caveats discussed in Chapter 1). Recall that our characteristic of interest is modeled as a random variable X and that we assume that it has a particular distribution representing the population that we are interested in. For example, if we ask a randomly chosen voter about whether they agree with a specific policy or not, we can model this as $X \sim \mathsf{Bern}(p)$. So, we record a 1 if the voter agrees and a 0 if (s)he doesn't. The fraction of voters in the population who agree is p. If we choose one voter at random, what do we expect this voter to do, agree or disagree? With probability p the voter agrees and with probability $1 - p$ (s)he disagrees. The probability weighted average of this is $p \cdot 1 + (1 - p) \cdot 0 = p$. Such a probability weighted average is called the **expectation** of a random variable and is denoted by $\mathsf{E}(X)$. Note that the expectation of a random variable is a characteristic of the population and, thus, a parameter. Often (but not always!) this parameter is denoted by μ.

Another way to think of the expectation is closely linked to the interpretation of probability as a frequency. Suppose that I compute the average value of X over all members of the population. Then a fraction p agrees with the policy and a fraction $1 - p$ disagrees. So, the frequency-weighted average of X will be p. Note that here I use the term "frequency-weighted" rather than "probability-weighted." The former refers to a situation where I average over the entire population, whereas the latter refers to averaging over one randomly chosen voter.

Let's now try and generalise the intuition from this example to a definition that works for *all* random variables. First start with a discrete random variable. Recall that such a random variable can take countably many values. We shall denote these by the set $\{x_1, x_2, \dots\}$. In the example of the voter, the set of possible outcomes is $\{0, 1\}$. If we had asked not just one voter, but n voters and counted the number of voters who agreed, then we would have $X \sim \mathsf{Bin}(n, p)$, in which case the range of possible values for X is $\{0, 1, 2, \dots, n\}$. In the general case, each of these outcomes $\{x_1, x_2, \dots\}$ has a label: the first has label 1, the second label 2, etc. If we are given the probability mass function f_X, then we know that the probability of the k-th value being observed is $f_X(x_k)$. So, the probability-weighted average of all possible outcomes is obtained by multiplying each possible outcome by its probability and adding all these numbers. In mathematical terms:

$$\mathsf{E}(X) = f_X(x_1)x_1 + f_X(x_2)x_2 + \cdots = \sum_{k=1}^{\infty} x_k f_X(x_k).$$

$\int$ In order to generalise this definition to continuous random variables, let $\mathscr{R}$ denote the range of possible values of the random variable X, and let f_X denote the density function. Of course, we cannot write the expectation as a sum anymore. After all, supposing that $\mathscr{R} = [0, 1]$, how do you "sum over every number in $[0, 1]$?" But, fortunately, we have a way of "summing over a continuum:" computing an integral. So, all we do is replace the sum by an integral:

$$\mathsf{E}(X) = \int_{\mathscr{R}} x f_X(x) dx.$$

A few useful rules for the expectation operator[8] are given below.

1. The mean of a constant is the constant itself: $\mathsf{E}(a) = a$, for all $a \in \mathbb{R}$.

2. The mean of a function of X can also be computed as a sum/integral, i.e.,

---

[8]The expectation is an example of what mathematicians call an *operator*. An operator is, loosely speaking, a mathematical object that transforms one object (in this case random variables) into another (in this case a real number). A function is an operator: it maps numbers into numbers. The differential is an operator: it maps a function into its derivative.

if $g : \mathscr{R} \to \mathbb{R}$, then

$$E[g(X)] = \begin{cases} \sum_{k=1}^{\infty} g(x_k) f_X(x_k) & \text{if } X \text{ discrete} \\ \int_{\mathscr{R}} g(x) f_X(x) dx & \text{if } X \text{ continuous.} \end{cases}$$

3. The expectation of a linear function of X equals the expectation of the linear function of X: $E(aX + b) = aE(X) + b$, for all $a, b \in \mathbb{R}$.

4. If g is non-linear, then life is more complicated. However, if g is convex (concave), i.e., if $g'' \geq 0$ ($g'' \leq 0$), then we can say: $E[g(X)] \geq (\leq) g(E(X))$. This is called **Jensen's inequality**.

5. If $X_1, X_2, \ldots, X_n$ are n random variables, then $E\left(\sum_{i=1}^{n} X_i\right) = \sum_{i=1}^{n} E(X_i)$.

Example 2.4. Suppose that you draw n independent observations from the same distribution with mean μ. List these n random variables as $X_1, X_2, \ldots, X_n$. We can now compute the **sample mean** of these random variables by taking their average, which we shall denote by $\bar{X}$, i.e.,

$$\bar{X} := \frac{1}{n} \sum_{i=1}^{n} X_i.$$

We can use the above rules to compute the mean of the sample mean:

$$E(\bar{X}) = E\left(\frac{1}{n} \sum_{i=1}^{n} X_i\right) = \frac{1}{n} E\left(\sum_{i=1}^{n} X_i\right) = \frac{1}{n} \sum_{i=1}^{n} E(X_i) = \mu.$$

So, the mean of the sample mean is the same as the population mean. ◁

A random variable X with mass/density f_X has a **symmetric** distribution around its mean μ if, for all x, it holds that $f_X(\mu - x) = f_X(\mu + x)$. For a symmetric distribution, the mean and median are the same: $\mu = \text{med} X$. A distribution is **skewed** to the right if $\mu > \text{med} X$ and skewed to the left if $\mu < \text{med} X$. From Figure 2.1 it is clear that $\text{Bin}(4, 0.5)$ is symmetric, $\text{Bin}(4, 0.25)$ is skewed to the right, and $\text{Bin}(4, 0.75)$ is skewed to the left.

2.3.2 Higher order moments

The expectation $E(X)$ of a random variable is often called the *first moment* of a distribution. In general, the **k-th moment** of a distribution is defined as

$$E(X^k) = \begin{cases} \sum_{\ell=1}^{\infty} x_\ell^k f_X(x_\ell) & \text{if } X \text{ discrete} \\ \int_{\mathscr{R}} x^k f_X(x) dx & \text{if } X \text{ continuous.} \end{cases}$$

These higher order moments are used as indicators for other aspects of a distribution. As we will see below, the second moment is used to provide a measure of *dispersion* around the mean. The third moment is used to provide a measure of *skewness* and the fourth moment gives a measure of *kurtosis* (fatness of tails). Skewness and kurtosis will not be further discussed here, but dispersion will be crucial to many of the procedures and methods discussed here.

It is very important to know, though, that not all moments of every distribution are finite. Some distributions even have an infinite or undefined mean.

2.3.3   Variance

If you want to measure the degree of dispersion of a distribution, then one thing you could do is compute how far, on average, possible outcomes are located from the mean. If this number is large, there is a lot of dispersion; if it is small, the possible values are clustered around the mean. Suppose that X is discrete. Then for each possible outcome we could measure the distance to the mean and then take the probability weighted average, i.e.,

$$\sum_{k=1}^{\infty}[x_k - \mathsf{E}(X)]f_X(x_k).$$

But now we have a problem: suppose that X has possible outcomes $\{-1, 0, 1\}$, with equal probabilities. Then $\mathsf{E}(X) = 0$ and

$$\sum_{k=1}^{3}[x_k - \mathsf{E}(X)]f_X(x_k) = \frac{1}{3}[(-1 - 0) + (0 - 0) + (1 - 0)] = 0.$$

This gives the impression that there is no dispersion around the mean, even though there clearly is. In order to solve this problem, we could *square* the distances to make all deviations positive. Doing so gives us the **variance**:

$$\mathsf{Var}(X) = \begin{cases} \sum_{k=1}^{\infty}[x_k - \mathsf{E}(X)]^2 f_X(x_k) & \text{if } X \text{ discrete} \\ \int_{\mathscr{R}}[x - \mathsf{E}(X)]^2 f_X(x)dx & \text{if } X \text{ continuous.} \end{cases}$$

The variance is often denoted by the symbol σ^2. Note that $\mathsf{Var}(X)$ does not have the same unit of measurement as X, because all values have been squared. In order to rectify this, we can take the square root to obtain the **standard deviation**: $\sigma = \sqrt{\mathsf{Var}(X)}$.

Some rules for the variance are given below.

1. The variance can be computed using the first and second moments: $\mathsf{Var}(X) = \mathsf{E}(X^2) - \mathsf{E}(X)^2$.

2. The variance of a constant is zero: $\mathsf{Var}(a) = 0$, for any $a \in \mathbb{R}$.

3. The variance of a linear function of X can be found explicitly: $\mathsf{Var}(aX + b) = a^2\mathsf{Var}(X)$, for all $a, b \in \mathbb{R}$.

4. The probability with which possible values of X can deviate from the mean is bounded. This result is known as the **Chebychev inequality**. Suppose that X is a random variable with mean μ and variance σ^2. Take any number $\varepsilon > 0$. Then

$$P(|X - \mu| \geq \varepsilon) \leq \frac{\sigma^2}{\varepsilon^2}.$$

2.4 Multivariate distributions

Just as a real number can be one-dimensional (a point on a line), two-dimensional (a point in the plane), three-dimensional (a point in the sphere), etc., so a random variable can be multi-dimensional. For example, when you test a new drug, you may wish to record both its effectiveness and its side effects and analyse them simultaneously. Just as a one dimensional random variable, a two-dimensional random variable $Z = (X, Y)$ has a distribution function, which is now a function of two variables:

$$F_Z(x, y) = P(\{X \leq x\} \cap \{Y \leq y\}).$$

Throughout this section I will assume that X and Y are continuous variables and that $Z = (X, Y)$ has a **joint density** f_Z. You can easily adjust what follows to the case of discrete random variables by replacing densities by point mass functions and integrals by sums. Suppose that the range of possible values of X and Y is given by $\mathscr{R}_x$ and $\mathscr{R}_y$, respectively. The **marginal density** of X is obtained by "removing" Y from the density f_Z. This is achieved by integrating over y, i.e.,

$$f_X(x) = \int_{\mathscr{R}_y} f_Z(x, y) dy.$$

This gives the density of X when viewed in isolation. In a similar way we get the marginal density of Y:

$$f_Y(y) = \int_{\mathscr{R}_x} f_Z(x, y) dx.$$

These densities can be integrated to give the distribution functions F_X and F_Y, respectively:[9]

$$F_X(x) = \int_{-\infty}^{x} f_X(s) ds, \quad \text{and} \quad F_Y(y) = \int_{-\infty}^{y} f_Y(s) ds.$$

---

[9]Note that x and y show up in the integration bounds and can therefore not be used as integration variables. I choose s instead.

Two random variables X and Y are **independent** if $F_Z(x, y) = F_X(x)F_Y(y)$. The **conditional distribution** of X given the event $\{Y \leq y\}$ is defined by

$$F_{X|y}(x) := \frac{F_Z(x, y)}{F_Y(y)}.$$

Note that this definition follows simply from the definition of conditional probabilities:

$$F_{X|y}(x) = \mathsf{P}(\{X \leq x\}|\{Y \leq y\}) = \frac{\mathsf{P}(\{X \leq x\} \cap \{Y \leq y\})}{\mathsf{P}(\{Y \leq y\})} = \frac{F_Z(x, y)}{F_Y(y)}.$$

So, two random variables are independent if

$$F_{X|y}(x) = F_X(x), \quad \text{for all } y \in R_y,$$

i.e, if my telling you something about the variable Y does not change your probability assessment of events related to X.

2.4.1   Association between two random variables

A measure of *association* between random variables is the **covariance**, which is defined as

$$\mathsf{Cov}(X, Y) = \mathsf{E}[(X - \mathsf{E}(X))(Y - \mathsf{E}(Y))].$$

A unit-free measure of association is the **correlation** coefficient, which is often denoted by ρ and is defined as

$$\rho = \frac{\mathsf{Cov}(X, Y)}{\sqrt{\mathsf{Var}(X)\mathsf{Var}(Y)}}.$$

Some properties of the covariance and correlation coefficient are listed below.

1. The following rule is often handy to compute covariances in practice: $\mathsf{Cov}(X, Y) = \mathsf{E}(XY) - \mathsf{E}(X)\mathsf{E}(Y)$.

2. The covariance is closely linked to the variance: $\mathsf{Cov}(X, X) = \mathsf{Var}(X)$.

3. The covariance of linear functions of random variables can be computed: $\mathsf{Cov}(aX + c, bY + d) = ab\mathsf{Cov}(X, Y)$.

4. Independent random variables have zero covariance: if X and Y are independent, then $\mathsf{Cov}(X, Y) = \rho = 0$.

5. The converse doesn't hold in general, apart from some special cases: if X and Y are jointly normally distributed and if $\mathsf{Cov}(X, Y) = 0$, then X and Y are independent.

6. The correlation coefficient measures association as a fraction: $\rho \in [-1, 1]$.

Example 2.5. Suppose that you draw n independent observations from the same distribution with mean μ and standard deviation σ. We already know that $\mathsf{E}(\bar{X}) = \mu$. Using the rules for covariance and variance, we can also compute the variance of the sample mean:

$$
\begin{aligned}
\mathsf{Var}(\bar{X}) &= \mathsf{Var}\left(\frac{1}{n}\sum_{i=1}^{n} X_i\right) \\
&= \frac{1}{n^2}\mathsf{Var}\left(\sum_{i=1}^{n} X_i\right) \\
&= \frac{1}{n^2}\left\{\sum_{i=1}^{n}\mathsf{Var}\left(X_i\right) + \sum_{j\neq i}\mathsf{Cov}\left(X_i, X_j\right)\right\} \\
&= \frac{1}{n^2}\sum_{i=1}^{n}\mathsf{Var}\left(X_i\right) = \frac{\sigma^2}{n}.
\end{aligned}
$$

So, the variance of the sample mean is a factor $1/n$ of the variance of the population. This means that the sample mean is clustered more closely around the mean μ than is the population itself. The more observations you have, the more clustered the sample mean is around the population mean. ◁

2.5 Normal distribution

The most important distribution in statistics is the **normal** distribution. If the possible values of a random value X are symmetrically distributed around a mean μ, the distribution has a "bell curve" shape, and outcomes far from the mean (on either side) become rarer and rarer in an exponential fashion, then X follows a normal distribution. This distribution has two parameters: the mean μ and variance σ^2. These two parameters determine the shape of the distribution. In fact, its density can be written as[10]

$$
f_X(x) = \frac{1}{\sigma\sqrt{2\pi}}e^{-\frac{1}{2}\left(\frac{x-\mu}{\sigma}\right)^2}.
$$

Note that this function indeed depends on only two parameters, μ and σ. If you draw the graph of this density, you get a symmetric bell-shaped curve around the mean μ. The parameter σ determines the spread around μ. We denote $X \sim \mathsf{N}(\mu, \sigma^2)$. A normally distributed random variable $Z \sim \mathsf{N}(0,1)$ is called **standard normal**. Throughout this book I reserve the letter Z for a standard normally distributed random variable.

---

[10]See Appendix G for properties of the exponential function.

The normal distribution has many appealing properties. Among them is the fact that computations involving the normal distribution are fairly straightforward. From the expression of the density function you may not expect this. In fact, the distribution function does not have an analytic expression! However, it turns out that all computations can be done once you know the values of the distribution function of the standard normal distribution. This distribution function is denoted by Φ, i.e.,

$$\Phi(z) := \mathsf{P}(Z \leq z) = \int_{-\infty}^{z} \frac{1}{\sqrt{2\pi}} e^{-s^2/(2\sigma^2)} ds.$$

Values of Φ are obtained from readily available tables and/or software. The central role played by the standard normal distribution comes from the fact that

$$X \sim \mathsf{N}(\mu, \sigma^2) \text{ implies that } Z := \frac{X - \mu}{\sigma} \sim \mathsf{N}(0, 1).$$

We call the procedure of subtracting the mean and dividing by the standard deviation **standardizing** the random variable X.

Example 2.6. Suppose that $X \sim \mathsf{N}(4, 9)$, i.e., $\mu = 4$ and $\sigma = 3$. Let's compute the probability that X exceeds 5. First, we can subtract the mean from both sides:
$$\mathsf{P}(X \geq 5) = \mathsf{P}\left(X - \mu \geq 5 - 4\right).$$

Now divide on both sides by σ:

$$\mathsf{P}(X \geq 5) = \mathsf{P}\left(\frac{X - \mu}{\sigma} \geq \frac{5 - 4}{3}\right).$$

The random variable on the right-hand side is standard normal and can be denoted by Z:[11]
$$\mathsf{P}(X \geq 5) = \mathsf{P}\left(Z \geq 0.33\right).$$

In order to write this so that we can use the distribution function Φ, we use property 2 of Proposition 2.1:

$$\mathsf{P}(X \geq 5) = 1 - \mathsf{P}\left(Z \leq 0.33\right) = 1 - \Phi(0.33).$$

Finally, we look up the value for Φ at 0.33 in a table:

$$\mathsf{P}(X \geq 5) = 1 - \Phi(0.33) = 1 - 0.6293 = 0.3707.$$

We could also compute the probability that X lies in the interval $[3, 5]$:

$$\mathsf{P}(3 \leq X \leq 5) = \mathsf{P}(-0.33 \leq Z \leq 0.33)$$
$$= \Phi(0.33) - \Phi(-0.33) = 0.6293 - 0.3707$$
$$= 0.2586.$$

---

[11]I typically round probabilities to four digits after the decimal point. That way you can easily read probabilities as percentages rounded to two digits. Non-probabilities are rounded to two digits, with intermediate computations rounded to three.

Note that you could also use the symmetry of the normal distribution:

$$\begin{aligned}
\mathsf{P}(3 \leq X \leq 5) &= \mathsf{P}(-0.33 \leq Z \leq 0.33) \\
&= 1 - 2\mathsf{P}(Z \geq 0.33) = 1 - (2)(0.3707) \\
&= 0.2586.
\end{aligned}$$

Finally, we often need to make **inverse probability calculations**. These are calculations of the type: "Find x such that $\mathsf{P}(X \geq x) = 0.05$." We execute these computations by turning the previous procedure upside down (inverting it). Start from the table of the standard normal and find a value z such that $\mathsf{P}(Z \geq z) = 0.025$. Using symmetry, this means we are looking for a value z such that $\mathsf{P}(Z \leq -z) = 0.025$. By looking in the table for the entry that gives $\Phi(-z) = 0.025$, we find that $z = 1.96$. We now need to "de-standardize" Z. Standardizing means subtracting μ and dividing by σ, so "de-standardizing" means doing things the other way around: multiply by σ and add μ. This gives

$$x = \mu + \sigma z = 4 + (3)(1.96) = 9.88.$$

We now know that $\mathsf{P}(X \geq 9.88) = 0.0250$. ◁

Recall from the previous sections that, if you have n independent draws from the same random variable with mean μ and variance σ^2, it holds that $\mathsf{E}(\bar{X}) = \mu$ and $\mathsf{Var}(\bar{X}) = \sigma^2/n$. If, for every $i = 1, \ldots, n$, it holds that $X_i \sim \mathsf{N}(\mu, \sigma^2)$, then it turns out that we can say even more: the shape of the distribution of $\bar{X}$ also turns out to be normal:

$$X_i \sim \mathsf{N}(\mu, \sigma^2), i = 1, \ldots, n, \text{ implies that } \bar{X} \sim \mathsf{N}(\mu, \sigma^2/n).$$

This is actually a special case of a more general result. Suppose that $X \sim \mathsf{N}(\mu_x, \sigma_x^2)$ and $Y \sim \mathsf{N}(\mu_y, \sigma_y^2)$, with X and Y not necessarily independent. Then

$$X \pm Y \sim \mathsf{N}(\mu_x \pm \mu_y, \sigma_x^2 + \sigma_y^2 \pm 2\mathsf{Cov}(X, Y)). \qquad (2.2)$$

2.5.1 Bivariate normal distribution

Suppose that you have two normally distributed random variables: $X \sim \mathsf{N}(\mu_x, \sigma_x^2)$ and $Y \sim \mathsf{N}(\mu_y, \sigma_y^2)$. Then it can be shown that $Z = (X, Y)$ is jointly normally distributed, with mean $\mu_Z = (\mu_y, \mu_y)$. The covariance between X and Y is denoted by σ_{xy}. The correlation then is $\rho = \frac{\sigma_{xy}}{\sigma_x \sigma_y}$. So, to fully describe the bivariate distribution of (X, Y), you need five parameters: $\mu_x, \mu_y, \sigma_x, \sigma_y$, and σ_{xy}. From the rules of covariance we know that, if X and Y are independent, then $\sigma_{xy} = 0$. For the bivariate normal distribution it turns out that it also works the other way around: if $\sigma_{xy} = 0$, then X and Y are independent.

In Chapter 9 we will work with the distribution of Y *conditional* on an observation x from the random variable X. We denote this random variable by $Y|\{X = x\}$. It turns out that

1. we can compute the conditional expectation, denoted by $\tilde{\mu}(x)$:

$$\tilde{\mu}(x) = \mathsf{E}(Y|X = x) = \mu_y + \frac{\sigma_{xy}}{\sigma_x^2}(x - \mu_x);$$

2. we can compute the conditional variance, denoted by $\tilde{\sigma}^2$:

$$\tilde{\sigma}^2 = \mathsf{Var}(Y|X = x) = \sigma_y^2 - \frac{\sigma_{xy}^2}{\sigma_x^2};$$

3. we can derive the distribution of $Y|\{X = x\}$:

$$Y|\{X = x\} \sim \mathsf{N}(\tilde{\mu}(x), \tilde{\sigma}^2).$$

It would take us way too far to prove all these results, so you can take them as given. Here's the main message: if Y and X are both normal random variables, then so is one conditional on the other. Yet another reason why the normal distribution is a "nice" distribution.

2.6   Limit theorems for the sample mean

Consider a random variable X with a mean $\mathsf{E}(X)$ and denote this mean by μ. The mean of the random variable is a parameter and can be interpreted as the average value of X in the population. In the real world we hardly ever observe the entire population but, instead, only observe a sample. Suppose now that we want to "guess" μ on the basis of this sample. Let's assume that each observation i can be seen as an independent draw X_i from the population, so that each X_i has the same distribution (and, in particular, the same mean μ). I suppose anyone would now compute the sample mean and use this as a guess for μ. So, if we have n observations, we compute $\bar{X} = \frac{1}{n}\sum_{i=1}^{n} X_i$. How do we know that this is a good approximation of μ?

There are several ways in which to answer this question and we will see some in later chapters. One intuition that most of us will have is that, if the sample is "very large," the sample mean will be "close to" the population mean. At this point mathematicians get very excited. What do you *mean* by "very large" and what do you *mean* by "close to?" One mathematical way of formalizing this is to say that we require that the *probability* of the sample mean deviating from μ by more than a tiny bit goes to zero as the sample becomes larger and larger. Formally this is written down as

$$\lim_{n\to\infty} \mathsf{P}(|\bar{X} - \mu| \geq \varepsilon) = 0, \quad \text{for all } \varepsilon > 0. \tag{2.3}$$

If this condition is satisfied, we say that $\bar{X}$ **converges in probability** to μ, and write $\bar{X} \overset{p}{\to} \mu$.

When looking at (2.3) you may wonder why all that complicated stuff about "for any $\varepsilon > 0$" is there. Why not simply require that

$$\lim_{n \to \infty} \mathsf{P}(\bar{X} = \mu) = 1?$$

This leads to a mathematical problem though. Suppose that X is a continuous random variable. Say that you draw a number at random between 0 and 1, so that $X \sim \mathsf{U}(0,1)$. What is the probability that you draw the number 0.5? Most people would say something like: "very close to zero." No. The probability is *exactly* 0.[12] But by the same token, what is the probability that $\bar{X}$ is ever *exactly* equal to μ, even if n gets very large? Precisely: zero. And that's why we write the statement (2.3) the way we do.

We can now prove that (2.3) actually holds.

Theorem 2.1 (law of large numbers (LLN)). *Suppose that $X_1, X_2, \ldots, X_n$ are independent and identically distributed random variables with $\mathsf{E}(X_i) = \mu$ and $\mathsf{Var}(X_i) = \sigma^2$, all $i = 1, 2, \ldots, X_n$. Then*

$$\bar{X} \overset{p}{\to} \mu, \quad as \ n \to \infty.$$

This theorem is actually not very difficult to prove and it nicely combines some of the properties we have encountered before.

Proof. We need to prove statement (2.3) for every possible $\varepsilon > 0$. So, let's pick any positive value for ε. Recall that

$$\mathsf{E}(\bar{X}) = \mu, \quad \text{and} \quad \mathsf{Var}(\bar{X}_n) = \frac{\sigma^2}{n}.$$

Applying Chebychev's inequality to $\bar{X}$ then gives that

$$\mathsf{P}(|\bar{X} - \mu| \geq \varepsilon) \leq \frac{\sigma^2}{n\varepsilon^2}.$$

As n goes to infinity, the right-hand side of this inequality goes to zero. The left-hand side is always non-negative because it is a probability (by Axiom (P1)). Therefore,

$$\mathsf{P}(|\bar{X} - \mu| \geq \varepsilon) \to 0, \quad as \ n \to \infty,$$

which proves the theorem. ■

Example 2.7. As an illustration of the LLN, consider the repeated tossing of a fair coin. The probability model for this is $X \sim \mathsf{Bern}(1/2)$. I simulated two sequences of 2,500 draws from this distribution. The evolution of the sample means is depicted in Figure 2.2. As is clearly visible, the sample mean converges to $1/2$, which is precisely the mean of the $\mathsf{Bern}(1/2)$ distribution.

---

[12]This has something to do with the fact that the interval $(0,1)$ is uncountable so that we can't speak meaningfully about "the smallest number larger than zero."

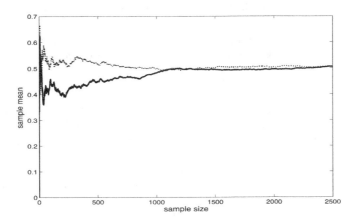

FIGURE 2.2: The evolution of the sample mean.

Suppose now that we do not know ex ante if the coin is fair. Our model is $X \sim \text{Bern}(p)$, where p is now an unknown probability of Heads. Suppose that the evolution of the sample mean looks like the solid line in Figure 2.3. After 1,500 coin flips it looks like $p = 1/3$. But what if we only had time to do 15 flips? I obtain (from reading the solid line) a sample mean of, roughly, 0.0667. Suppose that someone else had done the same experiment and observed the dotted line. This researcher, after 15 coin flips, would find $\bar{x} \approx 0.3477$. So, both researchers would come up with very different estimates of the same probability. However, for larger n the estimates get closer together. This is because of the LLN.   ◁

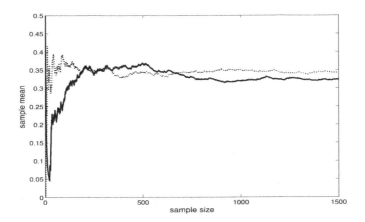

FIGURE 2.3: The evolution of the sample mean.

We now have a way in which to describe our intuitive notion that the sample mean converges to the population mean if the sample gets larger and larger. Another question we could ask is whether the *distribution* of the sample mean converges to some **limit distribution**. Just as in the case of the law of large numbers, we should ask ourselves what we mean by that precisely. That level of mathematical detail, however, will have to wait until a later date. At this stage we will content ourselves with a fairly imprecise statement of the so-called **central limit theorem**. This theorem gives us an approximate distribution of $\bar{X}$ for n large. I use the notation $\overset{A}{\sim}$ to indicate "is for large n approximately distributed as."

Theorem 2.2 (central limit theorem). *Suppose that you have independent random variables* $X_1, X_2, \ldots, X_n$, *all with the same distribution, with mean* μ *and variance* σ^2. *For large n it holds that* $\bar{X}$ *approximately (or asymptotically) follows a normal distribution. We write*

$$\bar{X} \overset{A}{\sim} N\left(\mu, \frac{\sigma^2}{n}\right).$$

A few comments on the CLT are appropriate.

1. The CLT only says something about the distribution of $\bar{X}$, *not* about the distribution of X itself. X can have any distribution, but as n becomes larger and larger, the distribution of $\bar{X}$ looks more and more like a normal distribution.

2. The CLT says that the distribution of $\bar{X}$ is approximately normal when n is large. That leads to two questions: (1) when is n large? and (2) how good is this approximation? Neither question can be answered in full generality. It all depends on the particular case. If, for example, $X \sim \text{Bern}(0.5)$, then the approximation is very good even for relatively small samples. If $X \sim \text{Bern}(0.25)$, the sample needs to be larger to provide a decent approximation. The example below illustrates this point.

Example 2.8. Suppose that you have n random variables $X_i \sim \text{Bern}(p)$, $i = 1, \ldots, n$. Then $\text{E}(X) = p$ and $\text{Var}(X) = p(1 - p)$ and, thus, $\text{E}(\bar{X}) = p$ and $\text{Var}(\bar{X}) = p(1 - p)/n$. Therefore, the CLT says that, for large n,

$$\bar{X} \overset{A}{\sim} N\left(p, \frac{p(1 - p)}{n}\right).$$

Standardizing this gives

$$Z = \frac{\bar{X} - p}{\sqrt{p(1 - p)/n}} \overset{A}{\sim} N(0, 1).$$

In order to make clear how Z depends on n, this is often rewritten as

$$Z = \sqrt{n}\frac{\bar{X} - p}{\sqrt{p(1 - p)}} \overset{A}{\sim} \mathsf{N}(0, 1).$$

So, for large n, the distribution of Z can be approximated by the standard normal.

How good is this approximation? Suppose $n = 12$ and $p = 0.05$. What is $P(\bar{X} \geq 3/25)$? In this case we know that $Y := \sum_{i=1}^{n} X_i \sim \mathsf{Bin}(n, p)$, so we can compute the exact probability. We get

$$P(\bar{X} \geq 3/12) = P(Y \geq 3) = 1 - P(Y = 0) - P(Y = 1) - P(Y = 2) = 0.02.$$

Using the normal approximation, we find

$$P(\bar{X} \geq 3/12) = P\left(Z \geq \sqrt{12}\frac{2.5/12 - 0.05}{0.218}\right) \approx 1 - \Phi(2.52) = 0.0059,$$

which is a poor approximation.

If, however, $p = 0.5$, we find as exact and approximate probabilities

$$P(\bar{X} \geq 3/12) = P(Y \geq 3) = 0.9807, \quad \text{and}$$

$$P(\bar{X} \geq 3/12) = P\left(Z \geq \sqrt{12}\frac{2.5/12 - 0.5}{0.5}\right) \approx 1 - \Phi(-2.02) = 0.9783,$$

respectively. This approximation is much better. [In the above, take note where the approximations take place, indicated by $\approx$.] In general, the accuracy depends on both p (better when p is close to $1/2$) and the sample size n (better when n large).   ◁

2.7   Chapter summary

We introduced the notion of a probability model and studied some basic properties of (conditional) probability. We introduced random variables, their distributions and their moments. In particular, we studied the normal distribution. The law of large numbers and the central limit theorem have been stated and their use indicated.

2.8   Exercises and problems

Exercises

Exercise 2.1. Prove the remaining four properties of Proposition 2.1.

Exercise 2.2. Let $(\Omega, \mathscr{F}, \mathsf{P})$ be a probability model and take some event $B \in \mathscr{F}$ with the property that $\mathsf{P}(B) > 0$. Show that the mapping $A \mapsto \mathsf{P}(A|B)$ defined in (2.1) is a probability. [You do this by carefully checking that (P1)–(P3) are satisfied.]

Exercise 2.3. A vase contains four white and two black marbles. You draw two marbles at random without replacement. Compute the probability of

(a) $A = \{$both marbles are white$\}$.

(b) $B = \{$both marbles have the same colour$\}$.

(c) $C = \{$at least one of the marbles is white$\}$.

(d) $D = \{$the second marble is white$\}$.

Exercise 2.4. Five (six-sided) dice are rolled. What is the probability of a "full house?" [A full house is a configuration with three of one and two of another.]

Exercise 2.5. Two fair (six-sided) dice are rolled. What is the conditional probability that at least one shows a six given that the dice show different numbers?

Exercise 2.6. Suppose that $X \sim \mathsf{Bin}(10, p)$. Compute $\mathsf{P}(X = 5)$ when

(a) $p = 0.7$.

(b) $p = 0.3$.

Exercise 2.7. Suppose that $X \sim \mathsf{H}(10, N, 0.7)$. Compute $\mathsf{P}(X = 5)$ when

(a) $N = 20$.

(b) $N = 50$.

(c) Compare your answers with the probability computed in Exercise 2.6(a) and comment on the difference/similarity.

Exercise 2.8. Suppose that $X \sim \mathsf{N}(2, 25)$.

(a) Compute $\mathsf{P}(X \geq 4)$.

(b) Compute $\mathsf{P}(X < 0)$.

(c) Compute $\mathsf{P}(0 \leq X \leq 4)$.

(d) Find a value for c such that $\mathsf{P}(X \geq c) = 0.1$.

Exercise 2.9. Suppose that $X \sim \mathsf{Geo}(0.7)$. Use the table in Appendix A to compute

(a) $\mathsf{P}(X = 2)$.

(b) $\mathsf{P}(X = 4|X \geq 2)$.

(c) The geometric distribution is an example of a **memoryless** distribution. Such distributions have the property that, for every h and for every x, it holds that $\mathsf{P}(X \geq x + h|X \geq h) = \mathsf{P}(X \geq x)$. Given your answers to parts (a) and (b), how would you explain this property in words?

In the next exercise, skip part (a) if you don't know integral calculus.

Exercise 2.10. Suppose that $X \sim \mathsf{Exp}(0.5)$.

(a) Use the table in Appendix B to show that the distribution function of X is given by

$$F_X(x) = 1 - e^{-x/2}.$$

Use the result in part (a) to compute

(b) $\mathsf{P}(X \geq 2)$.

(b) $\mathsf{P}(X \geq 4|X \geq 2)$.

(c) Is the exponential distribution memoryless?

In the next exercise, you can do part (a) even if you don't know integral calculus. Just draw a picture.

Exercise 2.11. Suppose that $X \sim \mathsf{U}(0, 3)$.

(a) Use the table in Appendix B to show that the distribution function of X is given by $F_X(x) = x/3$.

Use the result in part (a) to compute

(b) $\mathsf{P}(X \geq 1)$;

(b) $\mathsf{P}(X \geq 2|X \geq 1)$.

(c) Is the uniform distribution memoryless?

Exercise 2.12. Suppose that you have 100 independent draws $X_i \sim \mathsf{Bern}(0.8)$, $i = 1, \dots, 100$.

(a) Write down a normal approximation to the distribution of $\bar{X}$.

(b) Let $Z \sim \mathsf{N}(0, 1)$. What goes wrong when you use the approximation $\mathsf{P}(\bar{X} = 0.75) \approx \mathsf{P}(Z = -1.25)$?

(c) Argue that $\mathsf{P}(\bar{X} = 0.75) = \mathsf{P}\left(\sum_{i=1}^{n} X_i = 75\right) = \mathsf{P}\left(74.5 \leq \sum_{i=1^n} X_i < 75.5\right)$. [Note that there are no approximations here, just equalities.]

(d) Use the above observation to approximate $\mathsf{P}(\bar{X} = 0.75)$ using the normal distribution.

*In the above exercise you applied what is called a **continuity correction**. This correction is made necessary because you are approximating a discrete random variable by a continuous one.*

Exercise 2.13. Suppose that you have n independent draws $X_i \sim \text{Geo}(0.2)$, $i = 1, \ldots, n$.

(a) Use the table in Appendix A to compute $E(\bar{X})$ and $\text{Var}(\bar{X})$.

(b) For $n = 10$, compute $P(\bar{X} = 3)$.

(c) For $n = 100$, find an approximate distribution of $\bar{X}$.

(d) For $n = 100$, compute (approximately) $P(\bar{X} = 3)$.

Problems

Problem 2.1. Suppose that there are ten people in a group. What is the probability that at least two have a birthday in common?

Problem 2.2. Each morning a student rolls a die and starts studying if she throws 6. Otherwise, she stays in bed. However, during the four months of exams, she tosses a coin instead of a die and studies if she tosses heads. On a certain morning, the student is studying. What is the probability that it is exam period?

Problem 2.3. A particular disease affects 1 in 200 people. There exists a diagnostic test, which is 95% accurate (so, it correctly detects presence/absence of the disease in 95% of the cases). Your GP decides, randomly, to test you for the disease and it comes back positive. What is the probability that you have the disease?

Problem 2.4. You have been called to jury duty in a town where there are two taxi companies, Green Cabs Ltd. and Blue Taxi Inc. Blue Taxi uses cars painted blue; Green Cabs uses green cars. Green Cabs dominates the market with 85% of the taxis on the road. On a misty winter night a taxi killed a pedestrian and drove off. A witness says it was a blue cab. The witness is tested under conditions like those on the night of the accident, and 80% of the time she correctly reports the colour of the cab that is seen. That is, regardless of whether she is shown a blue or a green cab in misty evening light, she gets the colour right 80% of the time. What is the probability that a blue taxi caused the accident?

Problem 2.5. Suppose that the height (in cm) of fathers follows a normal distribution $X \sim N(168, 7^2)$ and that the distribution of sons' heights is $Y \sim N(170, 7^2)$. The correlation between a father's and son's height is $\rho = 0.4$.

(a) If a father is 5% taller than average, how tall, relative to the mean, do you expect the son to be?

(b) This phenomenon is called **regression to the mean**. Give an intuition for this terminology.

Problem 2.6. You are an investor in two stocks. If the price and time t of stock i is denoted by P_t^i and the dividend paid at time t is denoted by D_t^i, then the **return on stock** i at time t over the period $[t, t+1]$ is the random variable

$$R_t^i := \frac{P_{t+1}^i - P_t^i + D_t^i}{P_t^i}.$$

Suppose that you find that a good model for the log-returns is

$$\log(R_t^1) \sim \mathsf{N}(0.011, 0.037), \quad \text{and} \quad \log(R_t^2) \sim \mathsf{N}(0.0096, 0.079),$$

and that the correlation between the two log-returns is $\rho = 0.6$. If you hypothesize that the log-return of stock 2 will be 1%, what do you expect the log-return of stock 1 to be?

Problem 2.7 (based on Aitken et al., 2010). The *Sally Clark* case is one of the most important miscarriages of justice in the UK and revolves to a large extent around a probabilistic argument. Sally was convicted of killing two of her children. The argument that was used was based on research that showed that a child in a family like the Clarks' had a 1 in 8,543 chance of dying from sudden infant death syndrome (SIDS), which would be the explanation of the children's death in the case no foul play was involved. From this information an expert witness concluded that the probability of two SIDS deaths in the same family is 1 in 72,982,849, i.e., so small that SIDS could be ruled out as the cause of death. (In court, the expert likened this probability to the chances of backing an 80 to 1 outsider in the Grand National [horse race in the UK] four years running, and winning each time.)

(a) What assumption did the expert witness make about the event of individual children dying of SIDS in the same family? Is this a reasonable assumption?

The same study that was used to provide the 1 in 8,543 chance of SIDS in a family like the Clarks' also mentions that families with a history of SIDS have another SIDS case in 5 out of 323 cases.

(b) With this additional information, what is the probability of two siblings dying of SIDS in a family like the Clarks'? Was the conviction of Sally Clark on the basis of the expert witness's evidence correct?

Sally Clark was convicted in 1999 and had her conviction overturned in 2003, partly as a result of a campaign of the Royal Statistical Society. She died in 2007.

Problem 2.8 (based on Aitken et al., 2010). In the case *R v Deen* the following discussion (in relation to a DNA profile with a frequency of 1 in 3 million in the relevant population) took place between an expert witness and the prosecutor.

Prosecuting counsel: So the likelihood of this being any other man but [the defendant] is 1 in 3 million?

Expert: In 3 million, yes.

Prosecuting counsel: You are a scientist [...] doing this research. At the end of this appeal a jury are going to be asked whether they are sure that it is [the defendant] who committed this particular rape in relation to Miss W. On the figure which you have established according to your research, the possibility of it being anybody else being 1 in 3 million, what is your conclusion?

Expert: My conclusion is that the [sample] originated from [the defendant].

Prosecuting counsel: Are you sure of that?

Expert: Yes.

(a) Why is the expert convinced of the defendant's guilt? What probabilistic inference have they in mind?

(b) Is this probability correct?

(c) Do you have enough information to compute the probability of the defendant's guilt?

Suppose that there are 6 million a priori equally likely potential suspects.

(d) What is the probability of the defendant being guilty, taking into account the expert's evidence?

The mistake that the prosecuting counsel makes here is known in the literature as the **prosecutor's fallacy**.

Problem 2.9 (based on Sinn, 2009). After the 2007/08 financial crisis, credit rating agencies were criticized for not accurately reflecting credit risk; "securitization" was often mentioned as one of the root causes of the collapse of the sub-prime mortgage market. Let's try and understand the issues in a simple example. Suppose that we have a pool of four mortgages that are each paid back in full with probability .9 and that default with probability .1. A security is ranked AAA if its default probability is lower than 5%. (So, none of these mortgages is AAA.) Now we "securitize" the mortgages in the following way. We pool two mortgages together and divide this pool into a *senior tranche* and an *equity tranche*. The senior tranche defaults only if both mortgages default, whereas the equity tranche defaults when at least one mortgage defaults. This is a simple example of a *colleterized debt security* (CDS).

(a) What are the probabilities of default of the senior and equity tranches? Is the senior tranche AAA? Is the equity tranche AAA?

We now have two pools of two mortgages each and, thus, two senior tranches and two equity tranches. Suppose that we pool the two equity tranches into a senior tranche (which defaults only if both equity tranches default) and an equity tranche. This is a CDS^2.

(b) What are the probabilities of default of the senior and equity tranches? Is the senior tranche AAA? Is the equity tranche AAA?

So, out of four non-AAA securities we have created three AAA and one "toxic" asset.

(c) What assumption did you make to compute the probabilities in part (b)? Is this an appropriate assumption?

Suppose that the probability of a mortgage defaulting, conditional on another mortgage defaulting, is 0.6.

(d) Is the senior tranche of a (first-round) CDS, as constructed in part (b), still AAA?

Problem 2.10 (∗)**.** At your university a bus stop is served by two independent bus lines going to the railway station. On these lines, buses come at regular intervals of five and ten minutes, respectively. You get on the first bus that arrives at the stop. What is the expected waiting time?

Chapter 3

From Probability to Statistics

We can now return to the example in Section 1.4 and answer some of the questions posed there. The amount of caffeine in a randomly chosen cup of coffee can be thought of as a random variable. Suppose that you know that the standard deviation of caffeine content is 4 mg ($\sigma = 4$). Recall that the company claims that the average caffeine content is 50 mg ($\mu = 50$). A probabilist will now, most likely, model the amount of caffeine in a particular cup of coffee as the realization of a random variable $X \sim \mathsf{N}(50, 16)$.

Question 3.1. Why do you think a normal distribution may be a good choice here?

What does it mean that a random variable has an $\mathsf{N}(50, 16)$ distribution? Suppose that I buy 100 cups of coffee and measure their caffeine content. Drawing a histogram of the relative frequencies of caffeine contents gives the left panel in Figure 3.1. If I smooth this histogram, I get the graph depicted in the right panel of Figure 3.1. This is the graph of the density function of

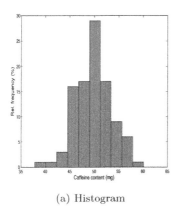
(a) Histogram

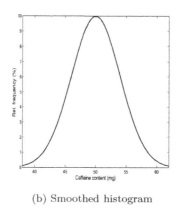
(b) Smoothed histogram

FIGURE 3.1: Caffeine content (mg) in 5,000 cups of coffee.

the $\mathsf{N}(50, 16)$ distribution. So, now I can answer a question like: "if I buy one cup of coffee, what is the probability that this cup will have a caffeine content between 45 mg and 55 mg?" The answer can be found using the tables of the normal distribution (after standardizing):

$$P(45 \leq X \leq 55) = P\left(\frac{45 - 50}{4} \leq \frac{X - \mu}{\sigma} \leq \frac{55 - 50}{4}\right)$$
$$= P(-1.25 \leq Z \leq 1.25) = \Phi(1.25) - \Phi(-1.25)$$
$$= 0.7888.$$

How should we *interpret* this probability? Many philosophers of science have thought about this and several interpretations have been developed. Throughout this book we will mainly use the so-called **frequentist** interpretation: the probability of an event is the relative frequency with which that event occurs in infinitely many repetitions of the experiment. This interpretation has some problems, as has been eloquently explained by Hacking (1965). I prefer to explain frequentist probability by appealing to a (hypothetical) universe consisting of all conceivable outcomes of the experiment. The probability of an event is then the frequency of all (hypothetical) realizations that satisfy the property stated in the event. So, in this interpretation the probability that *one randomly chosen* cup of coffee has a caffeine content between 45 mg and 55 mg is the same as the *frequency* of such cups in the entire population of all possible cups of coffee. In Appendix H I give a brief introduction to the **subjective** interpretation of probability, which views probability as a measure of personal (subjective) belief.

We can also compute the probability that you find a caffeine content of at least 52 mg in a randomly bought cup of coffee:

$$P(X \geq 52) = 1 - P(X \leq 52) = 1 - P\left(Z \leq \frac{52 - 50}{4}\right)$$
$$= 1 - \Phi(0.5) = 1 - 0.6915 = 0.3085.$$

Therefore, 30.85% of all possible cups of coffee have a caffeine content of at least 52 mg, or, alternatively, the probability that a randomly bought cup of coffee has a caffeine content of at least 52 mg is 30.85%. So, it shouldn't surprise you if the cup of coffee you actually bought has a caffeine content of 52 mg.

3.1   A first stab at statistical inference

The basic problem of statistical inference is the following: how do I *know* that the distribution of the caffeine content of cups of coffee is N(50, 16)? Well, in general you don't. So, a statistician will turn the analysis upside down: "suppose that I have information about the caffeine content of *some* cups of coffee from this coffee shop, what can I say about the distribution of the caffeine content of *all* cups of coffee?" The statistician then usually proceeds by fixing the *shape* of the distribution (for example, the normal) and, perhaps, even the standard deviation (there could be information about the coffee machine's precision from its technical specifications). The statistician then has a *model*: X follows an $N(\mu, 16)$ distribution. The task now is to "say something meaningful" about μ. Our greatest concern is the following: since we only have limited information (we are not observing *all* cups of coffee that could potentially be made with this machine), how do we deal with the uncertainty that this generates?

In our caffeine example you may wish to assess the coffee company's claim that the average caffeine content in its coffees is 50 mg. Suppose that you plan to buy nine cups of coffee. *Before* you buy these cups, the caffeine content for each cup is a random variable itself. Let's denote this random variable by X_i, for the i-th cup. We now have a *sequence* of random variables $(X_1, X_2, \ldots, X_9)$ that models our sample. We shall make two assumptions:

1. Each observation X_i is *independent* from all other observations.

2. Each observation X_i has the *same distribution* as X.

We call such a sample a **random**, or an **i.i.d.** (independent and identically distributed) sample, and I use both terms interchangeably.

Once you have conducted the experiment, suppose you find that the average caffeine content in your nine cups was $\bar{x} = 52$ mg.[1] Does this mean that we should doubt the claim of the manufacturer that the average content is 50 mg?

A naive answer would be to say: yes, because $52 > 50$. That would, however, be to ignore the fact that we only observe a sample and that each observation is subject to randomness. The statistician's answer is: let's first compute the *probability* that we observe an average caffeine content of at least 52 mg in a sample of nine cups *if the true average caffeine content in the population is 50 mg*. In order to do that we need to know the distribution of $\bar{X}$.

From probability theory we know that, if $X \sim N(\mu, \sigma^2)$, then

$$\bar{X} \sim N\left(\mu, \frac{\sigma^2}{n}\right).$$

So, if the the true mean caffeine content is 50 mg, then we know that $\bar{X} \sim N(50, 16/9)$. From this we can compute

$$P(\bar{X} \geq 52) = 1 - P(\bar{X} \leq 52) = 1 - P\left(\frac{\bar{X} - 50}{4/\sqrt{9}} \leq \frac{52 - 50}{4/3}\right)$$
$$= 1 - \Phi(1.5) = 0.0668.$$

---

[1]Since this number is computed *after* the caffeine content of each cup is analysed, this is a *realization* of the random variable $\bar{X}$. Therefore, it is denoted by a lower case letter.

This means that *if the true mean is 50 mg*, then the probability that we find a sample with a mean caffeine content of more than 52 mg is 6.68%. In the frequentist interpretation of probability: in the universe of all possible i.i.d. samples of nine cups of coffee with a mean caffeine content of 50 mg, the fraction of samples with a mean caffeine content of more than 52 mg is 6.68%.

Question 3.2. Do you think that, therefore, we have observed something that is so unlikely to occur that we should reject the claim that the average caffeine content is 50 mg? What would you consider a probability below which you find the company's claim effectively refuted?

3.2 Sampling distributions

If we are interested in the mean of a population, then we will often use the sample mean $\bar{X}$ as a "good guess." The sample mean is a quantity that can be computed after observing the data. Any such quantity is called a **statistic**. Note that, before you look at the data, a statistic is a random variable. Any observed value after plugging in the data is a realization. The distribution of such a statistic models the uncertainty that we face by observing the sample rather than the population.

In order to make probability statements of the kind we made in the previous section, we need to know the distribution of the statistic $\bar{X}$. Suppose that $(X_1, \ldots, X_n)$ is a random sample from a random variable $X \sim \mathsf{N}(\mu, \sigma^2)$. [I will often use the shorthand notation $(X_i)_{i=1}^n \overset{iid}{\sim} \mathsf{N}(\mu, \sigma^2)$ for this statement.] Then we know that

$$\bar{X} \sim \mathsf{N}\left(\mu, \frac{\sigma^2}{n}\right).$$

This distribution is called the **sampling distribution** of $\bar{X}$.

At this point, recall the "two worlds" analogy in Chapter 1: statisticians consider two worlds, the "real world" and the "sample world". The model with which we start, $X \sim \mathsf{N}(\mu, \sigma^2)$ with σ known, is the one we choose to describe the real world. In the coffee example, we assume that the caffeine content of all cups of coffee that could conceivably be bought has a relative frequency distribution that looks like a bell curve. In order to say something about this unknown mean, μ, we compute the sample mean of a sample of n, randomly chosen cups of coffee, $\bar{X}$. This sample mean "lives" in the sample world and it has a distribution, the sampling distribution. For our *particular* model of the real world, $X \sim \mathsf{N}(\mu, \sigma^2)$, the distribution in the sample world turns out to be $\bar{X} \sim \mathsf{N}(\mu, \sigma^2/n)$. Note how the distribution in the sample world depends on the model of the real world. Different models lead to different sample distributions.

As another example, suppose that $(X_1, \ldots, X_n)$ is a random sample from a random variable $X \sim \mathsf{Bern}(p)$. Since $\mathsf{E}(X) = p$, the mean of this population is p. We see that

$$\bar{X} = \frac{1}{n} \sum_{i=1}^{n} X_i = \frac{\# \text{ observations with } X_i = 1}{n},$$

which is the **sample proportion**. This, too, is a statistic. We will use the sample proportion (which I denote by $\hat{p}$) to say something about the population proportion p. Do we know the sampling distribution of $\hat{p}$? No, but we do know the distribution of $\sum_{i=1}^{n} X_i$, which is basically the same thing as $\bar{X}$:[2]

$$\sum_{i=1}^{n} X_i \sim \mathsf{Bin}(n, p).$$

Suppose that n is very large. Since $\mathsf{Var}(X) = p(1 - p)$ and $\mathsf{Var}(\hat{p}) = p(1 - p)/n$, we now know from the central limit theorem (CLT) that

$$\hat{p} \overset{A}{\sim} \mathsf{N} \left(p, \frac{p(1 - p)}{n} \right).$$

So, for n large we know what is called the **asymptotic sampling distribution** of $\bar{X}$.

Note that in both cases the variance of the sample mean is smaller than the variance of an individual observation by a factor $1/n$. So, the larger the sample, the smaller the variance of $\bar{X}$. Since the mean of $\bar{X}$ is the same as the mean of X, this simply says: the larger the sample, the less spread there is of $\bar{X}$ around the population mean. In statistical jargon: the larger the sample, the smaller the **sampling uncertainty**. This simple observation drives most of inferential statistics.

3.3 Chapter summary

Values of a variable in a population are modeled by random variables. A random sample is taken from this distribution and we aim to use it to find out something useful about the unknown parameter(s). Statistical inference will then be based on the probabilistic properties of an appropriate summary of the data.

---

[2]Recall Example 2.2.

3.4   Exercises and problems

Exercises

Exercise 3.1 (What's wrong?). Explain what is wrong in each of the following scenarios.

(a) If you roll a die three times and a one appears each time, then the next roll is more likely not to give a one.

(b) Suppose that $(X_i)_{i=1}^n \overset{iid}{\sim} \mathsf{Bern}(p)$. The sample proportion $\hat{p}$ is one of the parameters of the distribution of $\sum_{i=1}^n X_i$.

(c) Suppose that $X \sim \mathsf{Bin}(n, p)$. Then X represents a proportion.

(d) The normal approximation to the binomial distribution is always accurate when $n \geq 100$.

(e) When taking random samples from a population, the sample mean for larger sample sizes will have larger standard deviations.

(f) The mean of the sample mean changes when the sample size changes.

Exercise 3.2 (Modeling). In the following scenarios, indicate (i) the population, (ii) the variable, (iii) the parameter, and (iv) the statistic.

(a) A researcher wishes to know the average duration of unemployment among recent university graduates. She bases her conclusions on the average of a random sample of 210 unemployed recent university graduates.

(b) A marketing officer wants to know the proportion of Vogue readers who would buy a new mascara. She bases her conclusions on the sample proportion of a random sample of 89 Vogue readers agreeing with the statement "I will use this new mascara."

(c) A quality control officer is concerned about the variability (as measured by the variance) of the diameter of a hinge used in the tail sections of Airbus aircraft of the A320 type. She bases her conclusion on the sample variance of a random sample of 121 hinges obtained from the sole manufacturer of these hinges.

Exercise 3.3 (Analysis). Let $(X_1, \ldots, X_n)$ be a random sample from a random variable X with mean μ and variance σ^2.

(a) Compute $\mathsf{E}(\bar{X})$ and show how the assumptions of a random sample are used.

(b) Compute $\mathsf{Var}(\bar{X})$ and show how the assumptions of a random sample are used.

Exercise 3.4 (Analysis). Let $(X_1, \ldots, X_n)$ be a random sample from a random variable X with mean and variance equal to θ.

(a) Compute $\mathsf{E}(\bar{X})$ and $\mathsf{Var}(\bar{X})$.

(b) Use the tables in Appendix A and B to find a distribution that has the property that its mean and variance are the same.

Exercise 3.5. Let $(X_1, \ldots, X_{10})$ be a random sample from a random variable $X \sim \mathsf{N}(10, 2.3)$.

(a) Compute $\mathsf{P}(X < 10.8)$ and $\mathsf{P}(\bar{X} < 10.8)$.

(b) Compute $\mathsf{P}(X > 9.2)$ and $\mathsf{P}(\bar{X} > 9.2)$.

(c) Compute $\mathsf{P}(9.2 < X < 10.8)$ and $\mathsf{P}(9.2 < \bar{X} < 10.8)$.

(d) Find c and d such that $\mathsf{P}(10-c < X < 10+c) = \mathsf{P}(10-d < \bar{X} < 10+d) = 0.95$.

The next exercise may seem somewhat tricky. In part (d) you will have to exercise some judgement.

Exercise 3.6. Let $(X_1, \ldots, X_{10})$ be a random sample from a random variable $X \sim \mathsf{Bern}(0.45)$.

(a) Compute $\mathsf{P}(X \leq 0.6)$ and $\mathsf{P}(\bar{X} \leq 0.6)$.

(b) Compute $\mathsf{P}(X > 0.4)$ and $\mathsf{P}(\bar{X} > 0.4)$.

(c) Compute $\mathsf{P}(0.4 \leq X \leq 0.6)$ and $\mathsf{P}(0.4 \leq \bar{X} \leq 0.6)$.

(d) Find c such that $\mathsf{P}(0.45 - c \leq \bar{X} \leq 0.55 + c) \approx 0.95$.

Exercise 3.7. Let $(X_1, \ldots, X_{100})$ be a random sample from a random variable $X \sim \mathsf{Bern}(0.45)$. Use the normal approximation to

(a) compute $\mathsf{P}(0.4 < \bar{X} \leq 0.6)$.

(b) find c such that $\mathsf{P}(0.55 - c < \bar{X} < 0.55 + c) \approx 0.95$.

(c) Comment on the differences with the results you found in Exercise 3.6.

Problems

Problem 3.1. The government is planning to cut the budget for early child intervention programmes aimed at preventing children from disadvantaged backgrounds from falling behind in their development. The government minister proposing the cuts argues that the average age at which children fall behind is 11 and that, therefore, the programme is not particularly effective.

A researcher wants to know what is the average age at which children from disadvantaged areas start to substantially fall behind their peers. In order to investigate this, she conducts an experiment where she gives $n = 27$ children from a disadvantaged area a standardized test over a number of years. She compares each child's test score with the average national score of the same test. She finds that the average age at which children from disadvantaged backgrounds start to fall behind is 10.3 years. You may assume that the standard deviation is $\sigma = 2.3$ years.

If the government is correct, what is the probability that the researcher finds an average age of *at most* 10.3 years? What do you conclude about the government's claim? Answer this question in the following steps.

(M) Construct an appropriate model.

(A) Compute the probability that the researcher finds an average age of *at most* 10.3 years.

(D) Discuss the result.

Problem 3.2. The City of York is holding a referendum about whether it should have an elected mayor. In order to predict the result, the city council commissions an opinion poll among 18 randomly selected people on the electoral register. It turns out that ten voters are against and eight are in favour. If, in reality, the numbers of those in favour and against are equal, what is the probability that, in a poll like the one conducted, ten or more people are against? What would be the answer if, in reality, 60% of voters are in favour?

Chapter 4

Statistical Inference for the Mean based on a Large Sample

In this chapter the main ideas of statistical inference will be introduced in the context of a simple model. Throughout this chapter we will be interested in the *mean* of a population only and we will assume that we have access to a sample that is large enough for the central limit theorem to give a good approximation to the distribution of the sample mean $\bar{X}$.

4.1 Simple statistical model for the mean

As we saw in the previous chapter, a statistical analysis starts with a model of the population we are interested in. This model takes the form of a random variable with some distribution depending on some unknown parameter(s). There are no general rules that tell you what distribution to choose; experience and intuition usually drive this choice. This is something we will come back to in Chapter 5.

In many cases, however, we don't need to know the distribution of X. Suppose you are observing some random variable X (say caffeine content in a cup of coffee) and you are interested in the mean of X. Let's give this unknown probability-weighted average a name and call it μ, i.e., $\mu = \mathsf{E}(X)$. If the standard deviation of X is finite, we might denote it by, for example, σ, i.e., $\sigma^2 = \mathsf{Var}(X)$. In the terminology of the "two worlds" analogy, this is our model for the real world. Suppose further that we are going to observe n i.i.d. random variables $(X_1, \ldots, X_n)$. This will constitute the sample world, which we link to the real world by assuming that each X_i has the same mean μ.

Statistical inference consists of tools and techniques to "say something useful" about μ, based on realizations $(x_1, \ldots, x_n)$ of $(X_1, \ldots, X_n)$. In Chapter 6 we detail general procedures to construct and judge "ways to guess" μ based on the sample. I hope you'll agree that, intuitively, the best way to produce one number that we think is our "best guess" for μ, based on our sample, is to compute the sample mean $\bar{X}$. Note that this quantity is observed in the sample world.

As we saw in the previous chapter, in order to make inferential statements we need to know the sampling distribution of $\bar{X}$. In general this is not an easy task. Fortunately, as long as we have a large sample, the central limit theorem (CLT) tells us that $\bar{X}$ approximately follows a normal distribution, *regardless of the distribution of* X:

$$\bar{X} \overset{A}{\sim} \mathsf{N}\left(\mu, \frac{\sigma^2}{n}\right), \quad \text{for } n \text{ large.}$$

Note, however, that this distribution depends on two unknown parameters, μ and σ. Since we are only interested in μ, we call σ a **nuisance parameter**. Fortunately, there is a mathematical result that allows us to replace the unknown population variance σ^2 by the **sample variance**,

$$\hat{\sigma}_X^2 = \frac{1}{n}\sum_{i=1}^{n}(X_i - \bar{X})^2,$$

which we can compute from our data. It turns out that this does not change the approximate distribution of $\bar{X}$:

$$\bar{X} \overset{A}{\sim} \mathsf{N}(\mu, \hat{\sigma}_X^2/n).$$

This then is our model for the sample world and we will it use it in the remainder of this section to make inferences about the real world.

So, as long as we are interested in the mean of a population on the basis of a *large* sample, we know the sampling distribution of $\bar{X}$ (approximately). The larger the sample, the better the approximation. When is a sample large? That question can't be answered unequivocally. Many textbooks use 30 observations as a cut-off rule. If you suspect that the distribution you are drawing from is fairly symmetric, then 30 observations will often do. For asymmetric distributions you will need a larger sample than that. If you think this sounds rather vague, then you are right: it is. Unfortunately, the mathematics can't tell you anything more precise, because the CLT is formulated for the case that n gets infinitely large. (See Section 2.6 for an example.)

If we are willing to assume that our sample is large enough for the CLT to give a good approximation to the sampling distribution of $\bar{X}$, we can now use this distribution to answer questions like the following.

Question 4.1. A coffee chain claims that the average caffeine content of a regular Americano is 50 mg. You purchase 38 cups and find that the average caffeine content in your sample is $\bar{x} = 50.8$ mg, with a standard deviation of $\hat{\sigma}_X = 2.6$ mg. How confident are you about this estimate?   ◁

Question 4.2. A coffee chain claims that the average caffeine content of a regular Americano is 50 mg. You purchase 38 cups and find that the average caffeine content in your sample is $\bar{x} = 50.8$ mg, with a standard deviation of $\hat{\sigma}_X = 2.6$ mg. Does this sample provide sufficient evidence against the coffee chain's claim?   ◁

4.2   Confidence intervals

The issue addressed in Question 4.1 is one of **estimation**. If you were asked to provide an estimate of the average caffeine content based on your sample, you would probably say 50.8 mg. However, because we know about sampling uncertainty, we realize that if we had seen a different sample we would have obtained a different estimate.

Question 4.3. Suppose you obtain two samples of 38 cups of coffee. What is the probability that $\bar{X}$ will be different in both samples?

Therefore, we would like to answer Question 4.1 in a slightly different way. We know the (approximate) distribution of $\bar{X}$, so we could construct an *interval* such that we are, say, 95% confident that the true mean lies within that interval. In order to construct such an interval, recall that

$$\bar{X} \overset{A}{\sim} \mathsf{N}(\mu, \hat{\sigma}_X^2/n) \iff Z = \frac{\bar{X} - \mu}{\hat{\sigma}_X/\sqrt{n}} \overset{A}{\sim} \mathsf{N}(0, 1).$$

Using the table for the standard normal distribution, it is now easily obtained that
$$0.95 = \mathsf{P}(-1.96 \leq Z \leq 1.96).$$
For a general value α, you can find a number z such that

$$\mathsf{P}(-z \leq Z \leq z) = 1 - \alpha.$$

This number must be such that

$$\mathsf{P}(Z > z) = \alpha/2,$$

or, equivalently, such that

$$\mathsf{P}(Z \leq z) = 1 - \alpha/2, \quad \text{i.e.,} \quad \Phi(z) = 1 - \alpha/2.$$

Any number that satisfies $\mathsf{P}(Z \leq z) = p$, for some pre-specified probability p, is called the p-th **percentile** and is denoted by z_p. So, the number we are looking for here is the $1 - \alpha/2$-th percentile, $z_{1-\alpha/2}$, i.e.,

$$\mathsf{P}(-z_{1-\alpha/2} \leq Z \leq z_{1-\alpha/2}) = 1 - \alpha. \tag{4.1}$$

For example, if $\alpha = 0.05$, then (check!) $z_{1-\alpha/2} = z_{0.975} = 1.96$. See also Figure 4.1.

Using some standard algebraic operations we can now rewrite this in such

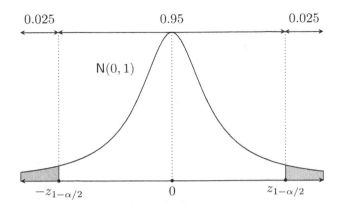

FIGURE 4.1: Tails of the standard normal distribution.

a way that we obtain an interval with μ "in the middle:"

$$1 - \alpha = \mathsf{P}\left(-z_{1-\alpha/2} \leq \frac{\bar{X} - \mu}{\hat{\sigma}_X/\sqrt{n}} \leq z_{1-\alpha/2}\right)$$

$$= \mathsf{P}\left(-z_{1-\alpha/2}\frac{\hat{\sigma}_X}{\sqrt{n}} \leq \bar{X} - \mu \leq z_{1-\alpha/2}\frac{\hat{\sigma}_X}{\sqrt{n}}\right)$$

$$= \mathsf{P}\left(\bar{X} - z_{1-\alpha/2}\frac{\hat{\sigma}_X}{\sqrt{n}} \leq \mu \leq \bar{X} + z_{1-\alpha/2}\frac{\hat{\sigma}_X}{\sqrt{n}}\right).$$

Question 4.4. Describe, in your own words, what this final equation means.

We now have an interval

$$\bar{X} \pm z_{1-\alpha/2}\frac{\hat{\sigma}_X}{\sqrt{n}}, \tag{4.2}$$

that guarantees that the true average μ lies within a fraction $1 - \alpha$ of all intervals constructed in this way. Note that this is a probability statement about the sample world, not the real world. In order to make a statement about the real world, for any particular interval thus constructed, the statistician now says that she "is $(1 - \alpha) \cdot 100\%$ confident that the true average lies in the interval."

You should be aware here of something very subtle: I did not use the word "probability." The reason is that in the *construction* of the interval I treat the sample mean as a random variable. *Before* I look at the result of my sample I can meaningfully speak about the probability that $\bar{X}$ falls in a particular interval. *After* I have looked at the results I have a *realization* of the sample mean, which is no longer a random variable. Therefore, I can no longer talk about probabilities. A simple example may illustrate. Consider a soccer player who is about to take a penalty. Before the penalty is taken I can say: "the probability that the player will score is 0.92." After the penalty is taken that statement is nonsense: the ball lies either in the goal or it does not. The only difference with statistics is that we never actually observe the location of the ball, as it were.

In any case, making a probability statement about a parameter makes no sense in the frequentist world, because the parameter is viewed as a constant, not a random variable. This is the main difference between frequentist and Bayesian statisticians: Bayesians *do* view parameters as (realizations of) random variables. See also Chapter 10.

An interval of the form (4.2) is called a **confidence interval**. In the case of (4.2) the **confidence level** is $1 - \alpha$. In principle you can take any confidence level you like, but in statistical practice the most commonly used levels are 90%, 95%, and 99% (i.e., $\alpha = 0.1$, $\alpha = 0.05$, and $\alpha = 0.01$, respectively). The only thing that changes in (4.2) when you change α is the percentile $z_{1-\alpha/2}$. See also Exercise 4.2.

Solution 4.1 (Solution to Question 4.1). Plugging all values into (4.2) we get the following 95% confidence interval for the average caffeine content: $(49.97, 51.63)$. This implies that we can report that we are 95% confident that the true average caffeine content in a regular Americano is between 49.97 mg and 51.63 mg. ◁

In Exercise 4.3 you will be asked to find what happens to the confidence interval in Equation (4.2) if any of its components change.

4.3 Hypothesis tests

Now that we know how to provide an interval of estimates for a parameter of interest, we can turn our attention to questions like those posed in Question 4.2. The way the question is phrased is reminiscent of a judicial trial: "is there enough evidence to find the accused guilty?" In fact, this analogy is a useful one to keep in mind while we think about what statisticians call **hypothesis tests**.

At the heart of such problems lies a statement about the true value of the average, called the **null hypothesis** and denoted by H_0. If you want to test the claim that the true μ is equal to some value μ_0, we write

$$H_0 : \mu = \mu_0.$$

In our trial analogy the null hypothesis is the claim "the defendant is innocent."

The opposing claim is called the **alternative hypothesis** and is denoted by H_1. Usually (but not exclusively) the alternative takes one of three forms:

$$H_1 : \mu > \mu_0, \quad H_1 : \mu < \mu_0, \quad \text{or} \quad H_1 : \mu \neq \mu_0.$$

The first two are called **one-sided** alternatives; the third one is called a **two-sided** alternative. The alternative can be thought of as the statement "the defendant is guilty."

4.3.1 The p-value

The judicial system prescribes that a defendant is to be found guilty when evidence of guilt is established *beyond reasonable doubt*. In statistics this is no different. First we need to think about what constitutes "evidence," then about what we mean by "beyond reasonable doubt." As evidence, statisticians often use a summary of the data that is called the p-**value**: the probability of obtaining at least the observed sample mean $\bar{x}$, or worse (in the direction of the alternative) conditional on the null hypothesis being true. So, the statistician starts with the assumption that H_0 is true (i.e., that the defendant is innocent) and then measures how far the evidence points away from the null hypothesis in the direction of the alternative (i.e., that the defendant is guilty).

In mathematical notation,

$$\text{pval} = \begin{cases} \mathsf{P}(\bar{X} \geq \bar{x} | \mu = \mu_0) & \text{if } H_1 : \mu > \mu_0 \\ \mathsf{P}(\bar{X} \leq \bar{x} | \mu = \mu_0) & \text{if } H_1 : \mu < \mu_0. \end{cases} \tag{4.3}$$

See Figure 4.2 for an illustration of the p-value for the case $H_1 : \mu > \mu_0$. As with confidence intervals, the p-value is a probability statement about the sample world, not the real world.

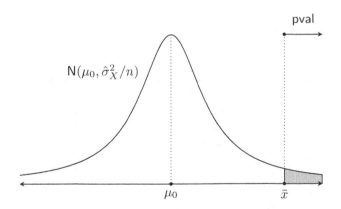

FIGURE 4.2: The p-value for $H_1 : \mu > \mu_0$.

Note that, since $\bar{X}$ follows a normal distribution (approximately), we will have to standardize $\bar{X}$, in order to compute the probability in (4.3). So, conditional on the null hypothesis being true, it holds that (approximately)

$\bar{X} \overset{A}{\sim} N(\mu_0, \hat{\sigma}_X^2/n)$. Standardizing then gives

$$Z = \frac{\bar{X} - \mu_0}{\hat{\sigma}_X/\sqrt{n}} \overset{A}{\sim} N(0, 1).$$

This implies that, if $H_1 : \mu > \mu_0$,

$$\text{pval} = P(\bar{X} \geq \bar{x}|\mu = \mu_0) = P\left(Z \geq \frac{\bar{x} - \mu_0}{\hat{\sigma}_X/\sqrt{n}}\right) = 1 - \Phi\left(\frac{\bar{x} - \mu_0}{\hat{\sigma}_X/\sqrt{n}}\right).$$

We can now also think about how to compute a p-value for the two-sided alternative $H_1 : \mu \neq \mu_0$, i.e., the probability of observing something more extreme than $\bar{x}$ conditional on H_0 being true. Transforming $\bar{X}$ to Z, this becomes: to compute the probability of something more extreme than $z := \sqrt{n}(\bar{x} - \mu_0)/\hat{\sigma}_X$, *in the direction of the alternative*. Since the alternative is two-sided, this means computing the probability

$$\text{pval} = \begin{cases} P\left(Z \leq -\frac{\bar{x} - \mu_0}{\hat{\sigma}_X/\sqrt{n}}\right) + P\left(Z \geq \frac{\bar{x} - \mu_0}{\hat{\sigma}_X/\sqrt{n}}\right) & \text{if } \frac{\bar{x} - \mu_0}{\hat{\sigma}_X/\sqrt{n}} > 0 \\ P\left(Z \leq \frac{\bar{x} - \mu_0}{\hat{\sigma}_X/\sqrt{n}}\right) + P\left(Z \geq -\frac{\bar{x} - \mu_0}{\hat{\sigma}_X/\sqrt{n}}\right) & \text{if } \frac{\bar{x} - \mu_0}{\hat{\sigma}_X/\sqrt{n}} < 0. \end{cases}$$

In shorthand notation, this can be written as

$$\text{pval} = 2P\left(Z \geq \left|\frac{\bar{x} - \mu_0}{\hat{\sigma}_X/\sqrt{n}}\right|\right).$$

Summarizing, the p-value is computed as follows:

$$\text{pval} = \begin{cases} P\left(Z \geq \frac{\bar{x} - \mu_0}{\hat{\sigma}_X/\sqrt{n}}\right) = 1 - \Phi\left(\frac{\bar{x} - \mu_0}{\hat{\sigma}_X/\sqrt{n}}\right) & \text{if } H_1 : \mu > \mu_0 \\ P\left(Z \leq \frac{\bar{x} - \mu_0}{\hat{\sigma}_X/\sqrt{n}}\right) = \Phi\left(\frac{\bar{x} - \mu_0}{\hat{\sigma}_X/\sqrt{n}}\right) & \text{if } H_1 : \mu < \mu_0 \quad (4.4) \\ 2P\left(Z \geq \left|\frac{\bar{x} - \mu_0}{\hat{\sigma}_X/\sqrt{n}}\right|\right) = 2\left(1 - \Phi\left(\left|\frac{\bar{x} - \mu_0}{\hat{\sigma}_X/\sqrt{n}}\right|\right)\right) & \text{if } H_1 : \mu \neq \mu_0 \end{cases}$$

Now that we have a way to measure evidence, all that is left to do is to decide what we call "beyond reasonable doubt." This standard needs to be chosen before we see the data (just like the goal posts are put on the pitch before a soccer match commences) and statisticians usually choose 1%, 5%, or 10% (note the analogy with confidence intervals). This standard (i.e., the probability beyond which we doubt the null) is called the **level** of the test and is usually denoted by α. So, if the p-value falls below α, then we conclude that we have observed something that is so unlikely to occur if the null hypothesis is true that we doubt it being true. Statisticians then say that they "reject the null hypothesis" or that "μ is **statistically significantly** different from μ_0 at level α." This is how frequentist statisticians translate a probability statement about the sample world into an inferential statement about the real world.

Solution 4.2 (Solution to Question 4.2). We first need to determine the hypotheses for this problem. The null hypothesis is obviously $H_0 : \mu = 50$. There is nothing in the problem formulation that suggests that one should care particularly about the direction in which the mean deviates from 50. Therefore, we choose $H_1 : \mu \neq 50$. The cut-off value we use is 5%.

Using the values given in Question 4.2, the p-value is easily computed as

$$\text{pval} = 2P\left(Z \geq \left|\frac{\bar{x} - \mu_0}{\hat{\sigma}_X/\sqrt{n}}\right|\right) = 2P\left(Z \geq \left|\frac{50.8 - 50}{2.6/\sqrt{38}}\right|\right)$$
$$= 2P(Z \geq 1.90) = 2(1 - \Phi(1.90)) = 0.0574.$$

So, on the basis of our sample, we cannot conclude that the average caffeine content is significantly different from 50 mg at the 5% level. ◁

4.3.2 Errors and power

As with all decisions in life, ours to reject H_0 may be wrong. There are two types of error that can be made: (i) we reject a null hypothesis that is, in fact, correct, or (ii) we fail to reject a null that is, in fact, wrong. These errors are called (rather unimaginatively) **Type I** and **Type II** errors, respectively. A Type I error is also called a **false positive** (because you are falsely claiming an effect that does not exist), whereas a Type II error is also known as a **false negative** (because you are falsely denying the presence of an existing effect). Of course, we never know whether we actually make an error at the time the decision is taken. We can, however, say something about the probability that an error is made, before a decision is taken.

In order to compute these probabilities, we need to change our focus a little bit. Recall that we reject H_0 if the p-value falls below the level α. In the case where the alternative is $H_1 : \mu > \mu_0$, this means we reject H_0 if

$$P(\bar{X} \geq \bar{x}|\mu = \mu_0) = P\left(Z \geq \frac{\bar{x} - \mu_0}{\hat{\sigma}_X/\sqrt{n}}\right) \leq \alpha.$$

We can also think of this in terms of rejecting on the basis of the observed realization $\bar{x}$ of $\bar{X}$ itself (recall our convention to denote random variables by upper case letters and their realizations by lower case letters). From the table for the standard normal distribution, it is easily found that we reject H_0 if

$$\frac{\bar{x} - \mu_0}{\hat{\sigma}_X/\sqrt{n}} \geq \Phi^{-1}(1 - \alpha) = z_{1-\alpha},$$

or, in terms of $\bar{x}$, if

$$\bar{x} \geq \mu_0 + z_{1-\alpha}\frac{\hat{\sigma}_X}{\sqrt{n}}. \tag{4.5}$$

The right-hand side of inequality (4.5) is called the **critical value** of the test and we will denote it by c (just to save on notation). In Exercise 4.4 you will find that, for alternative hypotheses $H_1 : \mu < \mu_0$ and $H_1 : \mu \neq \mu_0$, you reject H_0 if

$$\bar{x} \leq \mu_0 - z_{1-\alpha}\frac{\hat{\sigma}_X}{\sqrt{n}}, \quad \text{and} \tag{4.6}$$

$$\bar{x} \leq \mu_0 - z_{1-\alpha/2}\frac{\hat{\sigma}_X}{\sqrt{n}}, \quad \text{or} \quad \bar{x} \geq \mu_0 + z_{1-\alpha/2}\frac{\hat{\sigma}_X}{\sqrt{n}}, \tag{4.7}$$

respectively. The latter case is illustrated in Figure 4.3.

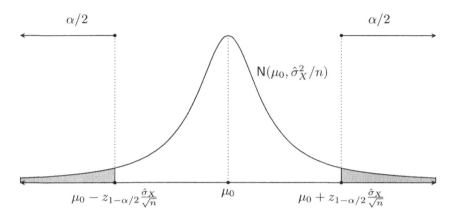

FIGURE 4.3: Rejection region for $H_1 : \mu \neq \mu_0$.

The probability of making a Type I error is the probability of rejecting H_0 if, in fact, H_0 is correct. That is, the probability that the sample mean exceeds the critical value if the null hypothesis is true. For $H_1 : \mu > \mu_0$, this probability is

$$\mathsf{P}(\bar{X} \geq c | \mu = \mu_0).$$

But this is, of course, exactly the level, α, of the test! So, by choosing our threshold α of "beyond all reasonable doubt" we have, in fact, fixed the probability of making a Type I error.

The probability of making a Type II error is denoted by β and is not so easily obtained. It is the probability of not rejecting the null H_0 if, in fact, H_0 is incorrect. So, you would think that this probability equals (for the alternative $H_1 : \mu > \mu_0$):

$$\mathsf{P}(\bar{X} < c | \mu > \mu_0).$$

This probability, however, cannot be computed because it does not fully specify the distribution of $\bar{X}$. Instead, it depends on the specific value of μ. Once we specify such a particular value for μ, say some $\mu_1 > \mu_0$, we can compute the error of making a Type II error:

$$\beta = \mathsf{P}(\bar{X} < c | \mu = \mu_1).$$

We can then compute the probability that we don't reject a correct null hypothesis:

$$Q = 1 - \beta.$$

This probability is called the **power** of the test. Of course, the power of the test will be different for every different value of μ_1. We will investigate this in an exercise.

We now come to an important point. We design tests by controlling for the probability of a Type I error; the probability of a Type II error then follows from this choice. We cannot simultaneously control for both error probabilities. In fact, by making α smaller we inadvertently make β larger. The probability β can actually be quite large, or, equivalently, the power can be quite low. This is why I've used the terms "rejecting" or "not rejecting" the null hypothesis. No serious statistician will ever use the phrase "accepting the null." The reason for this is that you might be using a procedure that doesn't reject the null hypothesis with a high probability even if H_1 is true. The way tests are designed is biased against rejecting null-hypotheses, just as the judicial system is biased against convicting innocent defendants. The flip-side of this is that the probability that we acquit guilty defendants can be quite high. Unfortunately, the only way to get that probability down is by increasing the probability of convicting innocents. You can see it in Figure 4.4: reducing the probability of a Type I error (the light gray area) requires increasing the critical value. But this increases the probability of a Type II error (the dark gray area). In statistics, you can't have your cake and eat it, unless you can increase the sample size.[1]

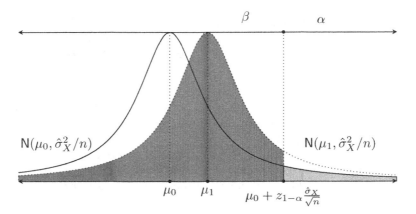

FIGURE 4.4: Probabilities of Type I and Type II errors.

Solution 4.3 (Solution to Question 4.2 continued). Let's compute the power

---

[1] See also Exercise 4.3.

of the test we designed against the particular alternative $H_1 : \mu = 51$. We find that the critical value of the test is

$$\mu_0 + z_{1-\alpha/2}\frac{\hat{\sigma}_X}{n} = 50 + 1.96\frac{2.6}{\sqrt{38}} = 50.827.$$

So, the probability of a Type II error is

$$\beta = P(\bar{X} < 50.827 | \mu = 51) = P\left(Z < \frac{50.826 - 51}{2.6/\sqrt{38}}\right) = \Phi(-0.41) = 0.3409.$$

This means that, although we only reject a false H_0 in 5% of tests thus constructed, we fail to reject a false H_0 in 34% of the cases. This is why you always have to be very careful equating "I fail to reject H_0" to "I accept H_0." The power of the test against the particular alternative $H_1 : \mu = 51$ is $Q = 1 - \beta = 0.6591$. ◁

4.4 Chapter summary

We have learned that we can base inference for the mean of a population on the basis of a large sample by appealing to the central limit theorem. We have provided examples of the two most common types of such inferences: confidence intervals and hypothesis tests.

4.5 Exercises and problems

Exercises

Exercise 4.1 (What's wrong?). Describe what is wrong for each of the statements below.

(a) Formula (4.2) can be used to compute a 95% confidence interval for any parameter.

(b) The size of a test gives the probability that, for a given sample, you wrongfully reject a correct null hypothesis.

(c) Formula (4.4) can be used to compute the p-value for testing the mean of a population based on a sample of 20 observations.

(d) In any test you can always make the size as small and the power as large as you want.

Exercise 4.2. The number 1.96 in Equation (4.2) depends on the confidence level, which was there chosen to be 95%. Find what this number will be if the confidence level is 90% or 99%.

Exercise 4.3. Describe for each of the cases below if the resulting confidence interval as given in Equation (4.2) gets wider or narrower. Explain your answer.

(a) The sample size, n, increases.

(b) The confidence level goes up.

(c) The standard deviation, $\hat{\sigma}_X$, increases.

Exercise 4.4. Show that the expressions in (4.6) are the critical values of a test of level α of the null hypothesis $H_0 : \mu = \mu_0$ against the alternatives $H_1 : \mu < \mu_0$ and $H_1 : \mu \neq \mu_0$, respectively.

Exercise 4.5. Let $(X_1, \ldots, X_{100})$ be i.i.d. random variables with mean μ and standard deviation $\sigma = 2$. Consider the hypothesis test $H_0 : \mu = 3$ vs. $H_1 : \mu < 3$.

(a) Compute the critical value of a test at the 5% level.

(b) Compute the power of this test against the alternatives $\mu = 2.5$, $\mu = 2$, and $\mu = 1.5$. Why does the power increase for lower values of the alternative?

Problems

Problem 4.1. A personnel manager has a random sample of 57 scores on an aptitude test that is given to applicants for summer internships. The sample mean score on the test is 67.9 with a (sample) standard deviation of 16.2 points. The firm makes recruitment decisions based on the assumption that the average score on the test is at least 70.

(M.i) What is the population in this situation?

(M.ii) What is the parameter that this sample information allows you to make inferences about?

(A.i) Construct a 95% confidence interval for this parameter.

(A.ii) Is the average score on the aptitude test significantly lower than 70 at the 5% level? At the 1% level?

(A.iii) What is the power of this test (at the 1% level) against the alternative hypothesis that the average score is 65?

(D) Advise the personnel manager on the use of this aptitude test for recruitment purposes.

Problem 4.2. You have a random sample of 77 prices paid for a particular gadget sold on an online auction site. The (sample) mean price is \$128.59 with a (sample) standard deviation of \$21.94.

(M.i) What is the population in this situation?

(M.ii) What is the parameter that this sample information allows you to make inferences about?

(A.i) Construct a 90% confidence interval for this parameter.

(A.ii) Is the average price for the gadget on the site significantly different from \$125 at the 10% level?

(A.iii) What is the power of this test against the alternative hypothesis that the average price is \$130?

Problem 4.3. A study compared various therapies for adolescents suffering from a particular eating disorder. For each patient, weight was measured before and after the therapy. For a particular therapy it was found that, in a random sample of 48 adolescents, the average weight gain was 6.54 pounds, with a standard deviation of 3.12 pounds. A therapy is deemed effective if weight gain exceeds 6 pounds.

(M.i) What is the population in this situation?

(M.ii) What is the parameter that this sample information allows you to make inferences about?

(A.i) Is the average weight gain in adolescents using the therapy significantly higher than 6 pounds at the 5% level?

(A.ii) What is the power of this test against the alternative hypothesis that the average weight gain is 7 pounds?

(D) Discuss the effectiveness of the therapy.

Confidence intervals and hypothesis tests are conducted given your sample. You can also use required statistical properties to determine the sample size that you need. The exercises below use familiar computations in a slightly different way, because their goal is slightly different.

Problem 4.4 (∗). Your client wants you to estimate the average length of a hospital stay in your region. From previous research it is known that the standard deviation is 2.4 days. Your client wants the estimate to be accurate with a margin of half a day either way, with 95% confidence. What sample size should you use? Break down your answer in the MAD components.

Problem 4.5 (∗). A lobby group wishes to establish if the average alcohol intake of UK men is significantly more than the government recommended limit of 4 units a day. From previous research it is known that the standard deviation of alcohol consumption is 1.83 units. The lobby group wants you to conduct an appropriate hypothesis test at the 1% level with a power of 80% against the alternative that average daily consumption is 4.5 units. What sample size should you use? Break down your answer in the MAD components.

Chapter 5

Statistical Models and Sampling Distributions

In this chapter we start a more general discussion of statistical inference. So far we have only been thinking about inference for the mean based on a large sample. But what if we are not interested in the mean but another parameter? Or what if we don't have a large sample? Throughout Chapter 4 we assumed that the sample mean is a good estimator for the population mean. But what do we mean by "estimator" and why is $\bar{X}$ a "good" estimator for $\mathsf{E}(X)$? All these questions will be answered in later chapters. Here we will think further about the basic building block for statistical inference: the *statistical model*.

5.1 Statistical models

Recall that in statistics we are interested in making inferences about a parameter (characteristic) of a population, based on a sample taken from the population. So, consider a population and a characteristic of this population that we are interested in. Statisticians start from the premise that the data they have available are the result of a **random experiment**. An experiment should have several characteristics, or desiderata. If results from an experiment are completely unpredictable (i.e., if there is no regularity at all), then we can't hope to derive any inferences from observations. Here then is our first desideratum.

(D1) Each trial's precise outcome is not known beforehand, but all possible outcomes are known and there is some regularity in the outcomes.

A second desideratum is that the experiment can be *replicated*.

(D2) The experiment can be repeated under identical conditions.

Example 5.1. A researcher wants to find out the proportion of voters who want to reduce defence spending. She proceeds by asking a random group of 53 voters for their opinion and records whether they agree with reducing defence spending or not.

This procedure satisfies (D1) and (D2) and can thus be called a random experiment:

(D1) The possible outcomes for each sampled voter are known a priori: "yes" or "no." For each sampled voter the researcher does not know a priori what the answer is going to be; if the fraction of all voters in favour of decreasing defence spending is p, then she knows that the probability of a single sampled voter being in favour is p.

(D2) The procedure can be repeated by any other researcher: anyone can sample 53 voters and ask the same question.   ◁

In this chapter we will try and capture the idea of a random experiment in a mathematically more precise way. The outcome of each sampled item is going to be modeled as the realization of a random variable. This random variable has some distribution function $F_\theta(\cdot)$,[1] where $\theta \in \Theta$ is a (finite-dimensional) vector of parameters from some set Θ, called the **parameter space**. In other words, the distribution of X is known up to a (finite) number of (unknown) parameters $\theta \in \Theta$. This is how we deal with (D1): it determines our probability model and in doing so specifies all possible outcomes and the chance regularity.

Example 5.2. Several commonly used models are listed below.

1. Suppose $X \sim \mathsf{N}(\mu, \sigma^2)$, with σ known. In this case the parameter is μ and the parameter space is $\mathbb{R}$.

2. Suppose $X \sim \mathsf{N}(\mu, \sigma^2)$, with both μ and σ unknown. The parameter is the (two-dimensional) vector (μ, σ) and the parameter space is $\mathbb{R} \times \mathbb{R}_+$. If μ is our only parameter of interest, then σ is called a **nuisance parameter**.

3. Suppose $X \sim \mathsf{Bern}(p)$. The parameter is the unknown success probability p and the parameter space is $(0, 1)$.   ◁

Condition (D2) states that the experiment must be *repeatable* under *identical conditions*. This is modeled in the following way.

Definition 5.1 (random sample). A *random* (or i.i.d.) *sample* of size n from a random variable X with distribution function $F_\theta(\cdot)$ is a sequence of random variables $(X_1, \ldots, X_n)$, with joint distribution[2]

$$F_\theta(x_1, \ldots, x_n) = \prod_{i=1}^{n} F_\theta(x_i). \qquad (5.1)$$

                                                                              ◁

Recall that (5.1) implies that we assume that all random variables $(X_i)_{i=1}^{n}$ are *independent* (repeatability) and that all are *identically distributed* (identical conditions). Hence the acronym "i.i.d." Note that, in Chapter 2, I denoted

---

[1] Recall that the distribution function completely determines the distribution.
[2] See Appendix F for the product notation.

distribution and density functions by F_X and f_X, respectively, to indicate their link to the random variable X. From now on we are in statistical territory, where the focus is on the unknown parameter θ. Therefore, I will now write F_θ and f_θ instead to draw attention to the dependence on the parameter.

Throughout this book it will be assumed that the data at your disposal are the result of a random (or i.i.d.) sample. How to *obtain* a random sample is a different matter indeed. How do you, for example, generate a representative poll, or a representative survey? And how do you deal with issues like non-response, or bias?[3] Statisticians have developed techniques to deal with such issues, but studying those would take us too far from our purpose here.

Another word of caution. The definition of random sample does *not* refer to the set of numbers you have in front of you if you do a statistical analysis. Any such sequence is a *realization* $(x_i)_{i=1}^n$ of the sequence of random variables $(X_i)_{i=1}^n$. There is a distinction between random variables (denoted by upper case letters) and realizations (denoted by lower case letters). The former are *functions* (from the sample space Ω to $\mathbb{R}$), whereas the latter are *numbers*. In the case of random variables, we can meaningfully assign probabilities and, therefore, talk about "the probability of X taking such and such a value". In the case of realizations we essentially observe nature's work *after* the probability P has been applied.

We are now ready to define exactly what the starting point of any statistical inference is, i.e., how we formalize the idea of a random experiment.

Definition 5.2 (statistical model). A *statistical model* consists of

(M1) a random variable X with a family of distributions $\{F_\theta(\cdot) : \theta \in \Theta\}$.

(M2) a random sample $(X_i)_{i=1}^n$ from X. ◁

I often use the following shorthand notation for a statistical model: $(X_i)_{i=1}^n \overset{iid}{\sim} F_\theta$. Note that a statistical model describes a model for the real world (in the two worlds analogy), but also provides a bridge to the sample world, by insisting that the sample is i.i.d.

Example 5.3. Let $(X_i)_{i=1}^n$ be an i.i.d. sample from $\mathsf{N}(\mu, 1)$, i.e., $(X_i)_{i=1}^n \overset{iid}{\sim} \mathsf{N}(\mu, 1)$. This means that we assume that all X_i are independent and have density

$$f_\mu(x) = \frac{1}{\sqrt{2\pi}} e^{-\frac{1}{2}(x-\mu)^2},$$

where μ is an unknown parameter. ◁

---

[3]Non-response is a serious issue in, for example, online opinion polling where many people are sent a request to participate, but only a few actually do. Bias is a problem in many applications using "big data," because the sample in such cases often comes from users of a particular technology. It is difficult to maintain that such is an i.i.d. sample from the population at large.

5.2   Some examples of statistical models

In this section we will look at some examples of statistical models. Several of these models will be analysed in greater detail in the chapters that follow.

Example 5.4. There is concern about the number of cases of misdiagnosis for some disease in a particular hospital. Say that a 1% rate of misdiagnosis is the internationally accepted norm for the disease in question. The population is the group of all (current and potential future) patients of the hospital. The characteristic of interest is the diagnosis status of a randomly chosen patient (misdiagnosed or not). Denote this variable by X. Let p denote the (unknown) probability that a randomly chosen patient is misdiagnosed. Then X can be modeled as a Bernoulli random variable with parameter p. There is, therefore, a problem with this hospital if $p > 0.01$. In order to investigate the hospital, you obtain a sample of size $n = 150$. Before you observe the sample, each observation is a Bernoulli random variable, $(X_1, \ldots, X_{150})$, with parameter p, i.e., $(X_i)_{i=1}^{150} \overset{iid}{\sim} \mathsf{Bern}(p)$ and the parameter of interest is the proportion p.   ◁

Example 5.5. You are the manager of a computer manufacturer who is responsible for the supply chain. Every week you get a shipment of 50 components. You wish to investigate the fraction of defective components, p. If this fraction exceeds 5%, you have an agreement with the supplier that you can send the shipment back. Investigating each component is far too time consuming and expensive. You therefore base your conclusions on a sample of five. Letting X denote the number of faulty items in a sample of 5 out of a shipment of 50, an appropriate model is $X \sim \mathsf{H}(5, 50, p)$. The parameter of interest is the proportion of faulty items, p.   ◁

Example 5.6. You manage a production process that fills bottles up to 1 liter. Because of legal considerations, you decide that the machine must be calibrated such that a sample of 10 bottles contains, on average, at least 1 liter with probability 0.95. From past experience you know that the machine has a standard deviation of 5 ml. Letting X denote the content of a randomly chosen bottle, an appropriate model is $(X_i)_{i=1}^{10} \overset{iid}{\sim} \mathsf{N}(\mu, 25)$. The parameter of interest is the mean content, μ.   ◁

Example 5.7. You manage a production process that fills bottles up to 1 liter. You are concerned about the variability of the content of bottles and you suspect that the standard deviation might be more than the 5 ml that you have been assuming for quality control purposes in the past. You will consider a sample of 20 bottles. Letting X denote the content of a randomly chosen bottle, an appropriate model is $(X_i)_{i=1}^{20} \overset{iid}{\sim} \mathsf{N}(\mu, \sigma^2)$. The parameter of interest is the standard deviation, σ.   ◁

Example 5.8. You are investigating the average caffeine consumption of citizens as part of a government-backed research project into the use of stimulants. If average daily consumption exceeds 150 mg, the government is considering starting a public awareness campaign. You have no idea what the distribution of caffeine consumption is, but you plan to get data from a random sample of 250 citizens and appeal to the CLT. Let X be the daily caffeine consumption of a randomly chosen citizen. Your model for the sampling distribution of $\bar{X}$ is $\bar{X} \overset{A}{\sim} \mathsf{N}(\mu, \hat{\sigma}_X^2)$, where $\hat{\sigma}_X$ is the sample standard deviation. The parameter of interest is the mean, μ. $\lhd$

Example 5.9. You are concerned about the number of complaints that your firm has received about the lifetime of the batteries it sells. You wish to ensure that at least 90% of your customers buy a product that lasts at least 100 hours. You investigate on the basis of a sample of 125. Letting X denote the lifetime of a randomly chosen battery, an appropriate model is $(X_i)_{i=1}^{125} \overset{iid}{\sim} \mathsf{Exp}(\lambda)$. The parameter of interest is the mean, λ. $\lhd$

Example 5.10. A pharmaceutical firm wishes to compare the average effectiveness of a new drug it has developed with the best drug currently available. A random sample of patients is randomly assigned to one of the two drugs. Say n_1 patients get the new drug, while n_2 patients get the existing drug. The measured effect is modeled by a random variable X_1 for the new and X_2 for the existing drug. Researchers suggest that an appropriate model is

$$(X_{1,i})_{i=1}^{n_1} \overset{iid}{\sim} \mathsf{N}(\mu_1, \sigma_1^2), \quad \text{and} \quad (X_{2,i})_{i=1}^{n_2} \overset{iid}{\sim} \mathsf{N}(\mu_2, \sigma_2^2).$$

The parameter of interest is $\mu_D := \mu_1 - \mu_2$. $\lhd$

5.3 Statistics

As stated before, the goal of inferential statistics is to use a random sample to say something about unknown parameters. We will want to do this by combining the information from the sample into a (several) number(s) that summarize(s) all the information that is available in the sample that is relevant for the unknown parameter(s). Such a summary is called a statistic.

Definition 5.3 (statistic). A *statistic* is a function $T(X_1, \ldots, X_n)$ of the sample, which does not depend on unknown parameters. $\lhd$

In the "two worlds" analogy, a statistic is the variable that is measured in the sample world.

Example 5.11. Let $(X_i)_{i=1}^n$ be a random sample from a random variable X. The following four often used quantities are statistics:

1. *sample mean*: $\bar{X} := \frac{1}{n} \sum_{i=1}^{n} X_i$.

2. *median*: in a sample with an odd number of observations, the median is the 0.5 percentile, $x_{0.5}$.

3. *sample variance*: $\hat{\sigma}_X^2 := \frac{1}{n} \sum_{i=1}^{n} (X_i - \bar{X})^2$.

4. *unbiased sample variance*: $S_X^2 := \frac{1}{n-1} \sum_{i=1}^{n} (X_i - \bar{X})^2 = \frac{n}{n-1} \hat{\sigma}_X^2$.

The reason for the term "unbiased" in the final statistic will become clear in Chapter 6. ◁

Note that, since statistics are functions of random variables, they are themselves random variables so that we can make probability statements about them. Most statistical inference is based on probability statements about carefully chosen statistics.

Statistics are based on the data, so it would be good to know if particular statistics can actually be used to say something about particular parameters. Since statistics are *summaries* of the data, it is inevitable that we lose some information captured in the data by looking at the statistic rather than the raw data. This is, of course, exactly the goal of statistical inference: summarizing the data in such a way that we keep the information pertaining to a particular parameter. Statistics that have this feature are called **sufficient**.

Definition 5.4 (sufficient statistic). Let $(X_i)_{i=1}^{n}$ be a random sample from a random variable X that has a distribution that depends on an unknown parameter θ, and let $T(\cdot)$ be a statistic. If the distribution of X conditional on $t = T(x_1, \dots, x_n)$ does not depend on θ, then $T(\cdot)$ is a *sufficient statistic* for θ. ◁

To state things in a simpler way: if we want to say something about a parameter θ and the statistic T is sufficient for θ, then reporting the realization of T gives you as much information about θ as reporting the entire sample. The following result is often used to check if a statistic is sufficient.

Theorem 5.1 (factorization criterion). *Let $(X_i)_{i=1}^{n}$ be a random sample with joint density (mass function) $f_\theta(x_1, \dots, x_n)$. Then $T(\cdot)$ is a sufficient statistic for θ if, and only if, there exist functions $g_\theta(\cdot)$ and $f(\cdot)$, such that*

$$f_\theta(x_1, \dots, x_n) = g_\theta(T(x_1, \dots, x_n))f(x_1, \dots, x_n).$$

Note that the function g_θ depends on the parameter θ and the (realization of the) statistic, whereas f depends on the sample, but not the parameter θ.

Example 5.12. Let $(X_i)_{i=1}^{n} \overset{iid}{\sim} \mathsf{Bern}(p)$. We show that $T = \bar{X}$ is a sufficient statistic for p:

$$f_p(x_1, \ldots, x_n) = \prod_{i=1}^{n} p^{x_i}(1-p)^{1-x_i} = p^{\sum_{i=1}^{n} x_i}(1-p)^{n-\sum_{i=1}^{n} x_i}$$

$$= \underbrace{p^{n\bar{x}}(1-p)^{n(1-\bar{x})}}_{g_p(\bar{x})} \times \underbrace{1}_{f(x_1,\ldots,x_n)} .$$

◁

Note that sufficient statistics are not necessarily unique. In the example above the statistic $\sum_{i=1}^{n} X_i$ is also sufficient.

Example 5.13. Let $(X_i)_{i=1}^{n} \overset{iid}{\sim} \mathsf{N}(\mu, \sigma^2)$, with σ known. We show that $\bar{X}$ is a sufficient statistic for μ:

$$f_\mu(x_1, \ldots, x_n) = \prod_{i=1}^{n} \frac{1}{\sigma\sqrt{2\pi}} e^{-\frac{1}{2\sigma^2}(x_i-\mu)^2}$$

$$= e^{-\frac{1}{2\sigma^2}\left(\sum_{i=1}^{n} x_i^2 - 2\mu \sum_{i=1}^{n} x_i - n\mu\right)} \left(\frac{1}{\sigma\sqrt{2\pi}}\right)^n$$

$$= \underbrace{e^{\frac{n\mu}{\sigma^2}(\bar{x}-1/2)}}_{g_\mu(\bar{x})} \underbrace{\left(\frac{1}{\sigma\sqrt{2\pi}}\right)^n e^{-\frac{1}{2\sigma^2}\sum_{i=1}^{n} x_i^2}}_{f(x_1,\ldots,x_n)}.$$

◁

Example 5.14. Let $(X_i)_{i=1}^{n} \overset{iid}{\sim} \mathsf{N}(\mu, \sigma^2)$, with μ and σ unknown. We show that $T = (\sum_{i=1}^{n} x_i, \sum_{i=1}^{n} x_i^2)$ is a sufficient statistic for (μ, σ^2):

$$f_{(\mu,\sigma^2)}(x_1, \ldots, x_n) = \prod_{i=1}^{n} \frac{1}{\sigma\sqrt{2\pi}} e^{-\frac{1}{2\sigma^2}(x_i-\mu)^2}$$

$$= \prod_{i=1}^{n} \frac{1}{\sigma\sqrt{2\pi}} e^{-\frac{1}{2\sigma^2}(x_i^2 - 2x_i\mu + \mu^2)}$$

$$= \left(\frac{1}{\sigma\sqrt{2\pi}}\right)^n e^{-\frac{1}{2\sigma^2}\left(\sum_{i=1}^{n} x_i^2 - 2\mu \sum_{i=1}^{n} x_i - n\mu\right)} \times 1$$

$$\equiv g_{(\mu,\sigma^2)}\left(\sum_{i=1}^{n} x_i, \sum_{i=1}^{n} x_i^2\right) f(x_1, \ldots, x_n).$$

◁

Example 5.15. Recall Example 5.10 with parameter of interest $\mu_D = \mu_1 - \mu_2$. From standard results for the normal distribution, we know the distributions of the sample mean effectiveness of the two drugs:

$$\bar{X}_1 \sim \mathsf{N}(\mu_1, \sigma_1^2/n_1), \quad \text{and} \quad \bar{X}_2 \sim \mathsf{N}(\mu_2, \sigma_2^2/n_2),$$

respectively. Since we assumed that patients were randomly assigned to each of the two groups, $\bar{X}_1$ and $\bar{X}_2$ are independent. This means that

$$\bar{X}_D := \bar{X}_1 - \bar{X}_2 \sim \mathsf{N}\left(\mu_D, \frac{\sigma_1^2}{n_1} + \frac{\sigma_2^2}{n_2}\right).$$

From our previous analysis this implies that $\bar{X}_D$ is a sufficient statistic for μ_D. ◁

5.4   Sampling distributions

Consider the statistical model $(X_i)_{i=1}^n \overset{iid}{\sim} \mathsf{N}(\mu, \sigma^2)$, with σ known. We know that $\bar{X}$ is a sufficient statistic for μ. As we saw in Chapter 3, we often need to make probability statements about (sufficient) statistics. That means we need to know their distributions. The distribution of a statistic is called its **sampling distribution**.

Definition 5.5 (sampling distribution). Let T be a statistic. The distribution of T is its *sampling distribution*. ◁

In the "two worlds" analogy, while the statistic describes our observation in the sample world, the sampling distribution describes the distribution in the sample world.

Example 5.16. Suppose $(X_i)_{i=1}^n \overset{iid}{\sim} \mathsf{Bern}(p)$ and consider the sufficient statistic $T = \sum_{i=1}^n X_i$. As we derived in Example 2.2, it holds that $T \sim \mathsf{Bin}(n, p)$. ◁

Example 5.17. Suppose $(X_i)_{i=1}^n \overset{iid}{\sim} \mathsf{N}(\mu, \sigma^2)$, with σ known and consider the statistic $T = \bar{X}$. It holds that $T \sim \mathsf{N}(\mu, \sigma^2/n)$. ◁

Example 5.18. Suppose $(X_i)_{i=1}^n \overset{iid}{\sim} \mathsf{N}(\mu, \sigma^2)$, with μ and σ unknown. Consider the statistic $T = \bar{X}$, which is sufficient for μ. In Appendix C.2 you can find that $\frac{T-\mu}{S_X/\sqrt{n}} \sim t_{n-1}$. ◁

Technically, the distribution in this last example is not a sampling distribution, since it is the distribution of a transformation of the statistic. But it is one that is very useful, as we will see in Chapter 7. Another example follows below.

Example 5.19. Suppose $(X_i)_{i=1}^n \overset{iid}{\sim} \mathsf{N}(\mu, \sigma^2)$. Consider the statistic $T = S_X^2$. In Appendix C.1 you can find that $(n-1)\frac{T}{\sigma^2} \sim \chi_{n-1}^2$. ◁

Example 5.20. Recall Example 5.8. We are interested in the parameter $\mu = \mathsf{E}(X)$, but did not make any assumptions on the distribution of X. However, from the CLT we know that $\bar{X} \overset{A}{\sim} \mathsf{N}(\mu, \hat{\sigma}_X^2/n)$. ◁

Example 5.21. Recall Example 5.15. The sampling distribution of $\bar{X}_D$ is $N(\mu_D, \sigma_1^2/n_1 + \sigma_2^2/n_2)$. Unfortunately, this distribution depends on the unknown nuisance parameters σ_1 and σ_2. If the samples n_1 and n_2 are large, we can use the CLT and replace the population variances with their sample analogues:

$$\bar{X}_D \overset{A}{\sim} N\left(\mu_D, \frac{\hat{\sigma}_1^2}{n_1} + \frac{\hat{\sigma}_2^2}{n_2}\right).$$

◁

Question 5.1. Let X denote the lifetime of a battery used in your production process. Suppose that it is known that $X \sim N(\mu, 25)$. The battery's supplier tells you that an average battery's lifetime is 70 hours. If this claim were true, what value should the average life of 25 batteries exceed with probability 0.95? If you observe $\bar{x} = 67$, what would you advise the manager?

Solution 5.1. If the claim is true, we have $E(\bar{X}) = 70$, $Var(\bar{X}) = 25/25$, and $\bar{X} \sim N(70, 1)$. We are looking for a number of hours, call it h, such that

$$P(\bar{X} > h) = 0.95.$$

Looking at Figure 5.1, we want to find h such that the grey area is 0.95.

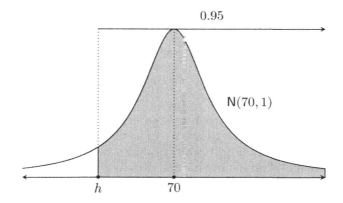

FIGURE 5.1: Inverse probability computation for Solution 5.1.

The formal computation is given below and should be familiar by now:

$$P(\bar{X} > h) = 0.95$$
$$\Longleftrightarrow P\left(\frac{\bar{X} - 70}{1} > \frac{h - 60}{1}\right) = 1 - P\left(\bar{X} - 70 \le h - 60\right) = 0.95$$
$$\Longleftrightarrow \Phi(h - 70) = 0.05$$
$$\Longleftrightarrow h - 70 = \Phi^{-1}(0.05) = -\Phi^{-1}(0.95) = -1.645$$
$$\Longleftrightarrow h \approx 68.36.$$

This suggests that we should reject the claim that $\mu = 70$. So, you should advise the manager to have a close look at the machine that produces the battery.   ◁

It is helpful to rephrase the essence of inferential thinking in the context of the "two worlds" analogy. If the average lifetime of batteries (in the real world) is 70 hours, then if I take a random sample of 25 such batteries (in the sample world), the probability that the average lifetime in the sample exceeds 68.36 hours is 95% (in the sample world). In other words, if I repeatedly take a random sample of 25 batteries from the sample world, then I should find an average lifetime of at least 68.36 hours in 95% of these samples. Given that I found an average lifetime of only 67 hours, I'm very confident that this is not due to "bad luck" in this particular sample. Therefore, I question the original claim (about the real world) that the average lifetime of a battery is 70 hours. All our conclusions are based on what would be *reasonable* to observe under certain conditions in hypothetically repeated samples. If our observations fall outside these bounds, then we question these conditions. To put it differently, (frequentist) inference uses observations from the sample world, and if these observations are unlikely, given our hypothesis about the real world, then we question our hypothesis about the real world. How this is done exactly will be the topic of Chapter 8. What will be good to remember though is that, in general, probability statements in frequentist statistics are about the sample world, not the real world.

With our more general terminology in hand, we can address new problems that do not involve the normal or Bernoulli distribution as well. This also shows that you don't have to have a great familiarity with a distribution to work with it.

Question 5.2. A factory is fitted with an emergency power unit with four spares. Let X denote the time to failure (in hours) of a power unit. So, X is a continuous random variable and can take values in $[0, \infty)$. From the table in Appendix B we find that the exponential distribution could be a good choice of distribution for X. This distribution depends on one parameter, λ. Note that $E(X) = \lambda$. Suppose that we know that the average lifetime of a power unit is 20 hours. Then $X \sim \text{Exp}(20)$. What is the probability that there is continuous power supply for at least a week?

Solution 5.2. We have five power units and the life-time of each follows an exponential distribution $\text{Exp}(\lambda)$, with $\lambda = 20$. So, the relative frequency distribution of the lifetime of a light bulb is the area under the graph of (see Appendix B)

$$f_{20}(x) = \frac{1}{20}e^{-x/20}, \quad x \geq 0.$$

The statistic that we need to say something about is

$$T(X_1, \ldots, X_5) = \sum_{i=1}^{5} X_i.$$

In fact, the probability that we need to compute is[4]

$$P\left(\sum_{i=1}^{5} X_i > 168\right).$$

In order to compute this probability, we need the sampling distribution of the random variable $\sum_{i=1}^{5} X_i$.

From the table in Appendix B we see that, if $X_i \sim \mathsf{Exp}(\lambda)$, then $\sum_{i=1}^{n} X_i \sim \mathsf{G}(n, \lambda)$. Denoting the distribution function of the gamma distribution by $F_{\mathsf{G}(n,\lambda)}$, we then get

$$P\left(\sum_{i=1}^{5} X_i > 168\right) = 1 - P\left(\sum_{i=1}^{5} X_i \leq 168\right) = 1 - F_{\mathsf{G}(5,20)}(168).$$

In practice, you use computer software to compute this probability. You can try to find the appropriate function in, for example, Excel.[5] This should give $P\left(\sum_{i=1}^{5} X_i > 168\right) = 0.0789$.
The remainder of the example deals with the computation of the above probability if you don't have a computer at hand.

We can also find this probability with the use of statistical tables.[6] For that we use Theorem C.1 in Appendix C, from which we conclude that

$$\frac{1}{10} \sum_{i=1}^{5} X_i \sim \chi_{10}^2.$$

Therefore,

$$P\left(\sum_{i=1}^{5} X_i > 168\right) = P\left(\frac{1}{10} \sum_{i=1}^{5} X_i > 16.8\right) = 1 - P\left(\frac{1}{10} \sum_{i=1}^{5} X_i \leq 16.8\right).$$

Table 3.2 in Neave's *Statistics Tables* gives the *inverse* distribution function of the χ_ν^2 distribution. So, in the row for $\nu = 10$ we look for the number that is closest to 16.8. This gives

$$P\left(\sum_{i=1}^{5} X_i > 168\right) = 1 - P\left(\frac{1}{10} \sum_{i=1}^{5} X_i \leq 16.8\right) \approx 1 - 0.925 = 0.075.$$

Note that this approximation is quite close to the exact solution. ◁

---

[4] Note that a week consists of 168 hours.
[5] In the toolbar: Formulas/Statistical/Gamma.Dist.
[6] I use Neave (1978).

5.5 Chapter summary

We formalized our notion of a statistical model and defined sufficient statistics as a tool to link statistics to parameters of interest. We also looked at sampling distributions of statistics.

Checklist for a successful statistical analysis

In this recurring section, we will gradually build a checklist for a proper statistical analysis. We can now add our first item.

1. Identify the population and the parameter of interest.

2. Identify the random variable, X, that is recorded during the experiment.

3. Choose an appropriate family of distributions for X and link the parameter(s) of interest to the parameter(s) of the family of distributions for X.

5.6 Exercises and problems

Exercises

Exercise 5.1 (What's wrong?). The following question appeared on an exam paper.

> In February 2002, the Associated Press quoted a survey of 3,000 UK residents conducted by YouGov.com. It stated that "only 21% wanted to see the monarchy abolished." You are asked to find if the proportion of UK residents in favour of abolishing the monarchy statistically significantly exceeds 20%. Formulate an appropriate statistical model for this situation. Motivate your choice.

A student answered as follows.

> It's a binomial $\mathsf{Bin}(np, nq)$, with $np = 630$ and $nq = 2370$.

What's wrong with this answer?

Exercise 5.2. Explain in your own words what a statistical model as defined in Definition 5.2 is. What does θ represent? Think of a real-world situation that you could encounter in your professional career, formulate a statistical model, and identify the parameter of interest.

Exercise 5.3. You conduct an opinion poll where you randomly select, with replacement, n members from a population of size $N > n$. The respondents can answer "agree" or "disagree" to the statement that you confront them with. Let X_i denote the random variable describing the i-th subject's answer.

(a) Does this constitute a random sample in the sense of Definition 5.2?

(b) Give the distribution of $\sum_{i=1}^n X_i$.

(c) What is the distribution of $\sum_{i=1}^n X_i$ as $N \to \infty$?

Exercise 5.4. You conduct an opinion poll where you randomly select, without replacement, n members from a population of size $N > n$. The respondents can answer "agree" or "disagree" to the statement that you confront them with. Let X_i denote the random variable describing the i-th subject's answer.

(a) Does this constitute a random sample in the sense of Definition 5.2?

(b) Give the distribution of X_i.

(c) Give the distribution of $\sum_{i=1}^n X_i$.

Exercise 5.5. Suppose that $(X_i)_{i=1}^n \overset{iid}{\sim} \mathsf{Bern}(p)$.

(a) What is the mass function of X_i?

(b) What is the mass function of the joint distribution of $(X_1, \ldots, X_n)$? Simplify as much as possible.

(c) If you had to "guess" p on the basis of a random sample $(X_1, \ldots, X_n)$, what would you do?

Exercise 5.6. Suppose that $(X_i)_{i=1}^n \overset{iid}{\sim} \mathsf{Poiss}(\lambda)$.

(a) Write down the mass function of the joint distribution of $(X_1, \ldots, X_n)$ and simplify as much as possible.

(b) Show that $\sum_{i=1}^n X_i$ is a sufficient statistic for λ.

(c) What is the sampling distribution of $\sum_{i=1}^n X_i$?

Exercise 5.7. Suppose that $(X_i)_{i=1}^n \overset{iid}{\sim} \mathsf{Exp}(\lambda)$.

(a) Write down the density function of the joint distribution of $(X_1, \ldots, X_n)$ and simplify as much as possible.

(b) Find a sufficient statistic for λ.

(c) What is the sampling distribution of this statistic?

Exercise 5.8 (∗). Suppose that $(X_i)_{i=1}^n \overset{iid}{\sim} \mathsf{U}(0,\theta)$, for some unknown $\theta > 0$.

(a) Write down the density function of the joint distribution of $(X_1,\ldots,X_n)$ and simplify as much as possible. (Hint: Be careful!)

(b) If you had to "guess" θ on the basis of the sample $(X_1,\ldots,X_n)$, what would you do? Can you derive the distribution of this statistic? (Don't be afraid to answer this with "no," but explain why.)

Problems

Problem 5.1. An investment bank has a target that none of its traders exceeds a certain threshold of what the firm calls "reasonable trading losses" on more than an average of two days per month. Management wishes to investigate whether its traders satisfy this criterion. In order to do so, it collects a random sample of $n = 10$ traders and records the number of days over the past month where the trader has a loss that exceeds the threshold. It is found that the average number of days that traders exceed the allowed loss is 2.8. The bank hires you to interpret this evidence.

(M.i) In this "experiment," what is recorded in each instance? Call this random variable X.

(M.ii) Is X a discrete or a continuous random variable?

(M.iii) Using the appendix, what do you think is an appropriate distribution for X?

(M.iv) Given your choice in (M.iii), are there any unrealistic assumptions that you have to make? If so, argue whether these assumptions seriously impede your analysis.

(M.v) What is the parameter of interest in this case?

(M.vi) Formulate your statistical model.

(A) Suppose that management is correct and that the average number of days a trader suffers a large loss is two per month. What is the probability that in a sample of 10 you find an average that is at least 2.8?

(D) Based on the probability computed in (A), how do you interpret this evidence and what do you report to management?

Problem 5.2. Many commercials and adverts for beauty products use a statistical claim. For example, in the TV advert for Maybelline's "Dream Matte Mousse" at some point it is stated that "82% of 169 agree." You can find the advert here: http://www.youtube.com/watch?v=KnDtsDOcAKA.

(a) What do 82% of 169 agree with?

(b) Do we know whether the 169 constitute a random sample?

(c) Do we know what exactly these 169 were asked?

(d) Does the information given allow us to perform a proper statistical analysis? Try and find more information on, for example, the internet.

The problems below are all exercises in formulating appropriate statistical models.

Problem 5.3. A bus company servicing a university advertises that its buses run at least every 10 minutes. A student is suspicious and decides to randomly show up at the bus stop n times. Construct an appropriate statistical model and clearly indicate what is the parameter in this case and its distribution. If you can, find a sufficient statistic for the parameter and its sampling distribution.

Problem 5.4. A consumer organization wishes to investigate a soft drink manufacturer's claim that its soft drinks contain, on average, 12% of an adult's GDA (guideline daily amount) of sugar per 100 ml. It will base its conclusion on sugar measurements from a random sample of 100 ml batches of the soft drink. Construct an appropriate statistical model and clearly indicate the parameter in this case. If you can, find a sufficient statistic for the parameter and its sampling distribution.

Problem 5.5. A city council is worried about the variability of student attainment in a particular standardized test. It wishes to investigate this based on a random sample of student scores on the test. Construct an appropriate statistical model and clearly indicate the parameter in this case. If you can, find a sufficient statistic for the parameter and its sampling distribution.

Problem 5.6. A government department is worried about the average length of time job seekers spend looking for a job. It plans to investigate the issue based on the amount of time a random sample of job seekers spends looking for a new job. Construct an appropriate statistical model and clearly indicate the parameter in this case. If you can, find a sufficient statistic for the parameter and its sampling distribution.

Problem 5.7. A firm is concerned about the fraction of faulty items it receives from a supplier. It intends to investigate this by taking a random sample of items and determining their status. Construct an appropriate statistical model and clearly indicate the parameter in this case. If you can, find a sufficient statistic for the parameter and its sampling distribution.

Problem 5.8. A firm produces razor blades. For quality control purposes, a manager wants to know the average lifetime of a blade, based on a sample. Lifetime here is defined as the first shave for which the blade fails to meet particular performance measures. Construct an appropriate statistical model and clearly indicate the parameter in this case. If you can, find a sufficient statistic for the parameter and its sampling distribution.

Problem 5.9. A think tank wants to know the fraction of CEOs among FTSE100 firms[7] who agree with the statement that "austerity is not the best way to bring growth back to the UK economy." It wants to base its conclusions on a survey among a randomly chosen number of FTSE100 CEOs ($n < 100$). In order to avoid embarrassment, it will only ask each CEO at most once. Construct an appropriate statistical model and clearly indicate the parameter in this case.

The next problem is essentially an exercise in using sampling distributions and performing probability calculations using software and/or statistical tables.

Problem 5.10. In an operating theatre it is of the utmost importance that there is a constant source of light. The hospital manager wants to know how many lamps with an expected life of 1,000 hours should be bought if the probability of continuous operation for two years has to equal 0.95. Only one lamp will burn at any one time and if the light fails, a new lamp is immediately installed. The light will operate continuously. How many lights should be bought?

---

[7]The FTSE100 is the main stock index on the London Stock Exchange, comprising of the 100 largest quoted companies.

Chapter 6

Estimation of Parameters

6.1 Introduction

In Exercises 5.5 and 5.8 you were asked to "guess" the parameter of a statistical model (p and θ, respectively) on the basis of a sample. In the case of Exercise 5.5 this was probably not very difficult. In Exercise 5.8, however, it is not so obvious what you should do.

In this chapter we look at the following question: *given a statistical model, what is a reasonable "guess" for the parameter?* We will discuss particular statistics (in the sense of Chapter 5), which we call **estimators**. For any realization of a random sample, these lead to realizations of the estimators, which we call **estimates**. It is important to note that estimators are random variables (on which we can make probability statements), whereas estimates are realizations of random variables. We will look at the following issues: given a statistical model $\{\, F_\theta \mid \theta \in \Theta \,\}$:

1. can we come up with *general* procedures to estimate θ? (Sections 6.2 and 6.4) and

2. can we judge the *quality* of estimators in a useful way? (Section 6.3).

6.2 Maximum likelihood estimators

Suppose that we are interested in a parameter θ which can take values in some set Θ. Let $(X_i)_{i=1}^n$ be a random sample from X with family of distributions $\{F_\theta(\cdot) : \theta \in \Theta\}$.

Definition 6.1 (estimator). An estimator for a parameter θ is a statistic $\hat\theta = \hat\theta(X_1, \ldots, X_n)$ taking values in the parameter space Θ. ◁

A notational convention is that an estimator for a parameter is denoted with a "hat." Note that estimators "live" in the sample world. The requirement that they take values in the parameter space ensures that there is a link between the sample world and the real world.

In this section we will think about the question: how do you construct estimators? We will look at a general principle that we can apply to any statistical model. The advantage of using general principles is that it prevents "ad hoc-ery." In addition, combined with what we will learn in the next section, it allows us to derive *general* results about properties of a large class of estimators.

In principle, any statistic could be called an estimator. Some of them would not make much intuitive sense, as the following example shows.

Example 6.1. You toss a (possibly biased) coin n times and model this as $(X_i)_{i=1}^n \overset{iid}{\sim} \mathsf{Bern}(p)$. Suppose that you want to estimate p. Here are two possible estimators:

$$\hat{p}_1 = X_1 \quad \text{and} \quad \hat{p}_2 = \bar{X}.$$

Both are statistics taking values in $\Theta = [0,1]$ and, therefore, valid estimators for p. [In fact, $\hat{p}_1$ only takes values in $\{0,1\}$.] It will be intuitively clear, however, that $\hat{p}_2$ is "better" than $\hat{p}_1$. ◁

There are several well-established methods of obtaining estimators. In this book we will mainly focus on the method of **maximum likelihood**.[1] The basic idea is as follows: we have a statistical model with some unknown parameter θ and "nature" has given us observations drawn from a distribution that depends on θ. Our best estimate for θ, then, is to find the value $\hat{\theta}$ that makes *what we have observed the most likely outcome*. This way of estimating uses the idea of **likelihood**.

Definition 6.2 (likelihood function). Let $(X_i)_{i=1}^n$ be a random sample from X with density (or mass) function $f_\theta(x)$. The *likelihood function* of $(X_i)_{i=1}^n$ is

$$L(\theta) := \prod_{i=1}^n f_\theta(x_i). \tag{6.1}$$

◁

The likelihood function is nothing more than the joint density (or mass) function of a random sample. The joint density function is a function of the observations given the parameter, but, because we are interested in the parameter and take the sample as given, we think about this joint density as a function of the parameter given the observations.

The maximum likelihood approach prescribes how to find the value of θ that maximizes the likelihood function. The idea behind it is that nature has chosen the realization $(x_i)_{i=1}^n$ of the random sample $(X_i)_{i=1}^n$ according to the likelihood function $L(\theta)$. It then makes sense to choose θ such that the likelihood of observing $(x_i)_{i=1}^n$ is maximal.

---

[1] You can see another method in Section 6.4.

Definition 6.3 (maximum likelihood estimator). The *maximum likelihood* (ML) *estimator* for θ, denoted by $\hat{\theta}_{ML}$, maximizes the likelihood function, i.e.,[2]

$$\hat{\theta}_{ML} = \arg\max_{\theta \in \Theta} L(\theta). \tag{6.2}$$

◁

In practice, solving the maximization problem (6.2) is often difficult to do by hand, so computer software is used instead. It is instructive, however, to see how it works in simple cases, but this does require some calculus. If you feel uncomfortable with this, feel free to take the results as given and read around the calculus.

Recall that the likelihood function is essentially the joint density function of the random sample (but thought of as a function of the parameter). It is thus the product of the density function of each observation. Differentiating this product then requires you to use the product rule many times, leading to an algebraic mess. There is, however, a way out. If you draw the graph of $f(x) = \log(x)$, then you see that this is a strictly increasing function. Suppose that you want to maximize the function

$$f(x) = e^{-x^2}, \quad \text{on } [0, \infty).$$

Verify for yourself that the maximum is attained at $x = 0$ and that the maximum is $f(0) = 1$. Now consider the function $g(x) = \log(f(x))$. Verify that the maximum is attained at $x = 0$ and that the maximum is $g(0) = 0$. So, we find that

$$\arg\max f(x) = \arg\max g(x), \quad \text{but} \quad \max f(x) \neq \max g(x).$$

This is a general property: any strictly monotone transformation of a function has the same maximum *location* but a different maximum *value*.

Here we want to maximize the likelihood function L. We are not interested in the maximum value itself, but in the value of θ that maximizes the likelihood. So, applying a strictly monotone transformation to L does not change this value of θ. Therefore, we often find the ML estimator by maximizing the **log-likelihood**.[3]

Definition 6.4 (log-likelihood). The *log-likelihood* is the (natural) logarithm of the likelihood function:

$$\ell(\theta) := \log(L(\theta)).$$

◁

---

[2]The notation arg max means that, rather than looking at the maximum *value* of the function, we look at the *argument* at which the maximum is obtained.

[3]See Appendix G for properties of the logarithmic function.

The log-likelihood function turns the product in (6.1) into a sum, which often makes the optimization problem easier to solve.

Question 6.1. Let $(X_i)_{i=1}^n \overset{iid}{\sim} \mathsf{Bern}(p)$. Find the ML estimator for p.

Solution 6.1. We first derive the likelihood and log-likelihood functions. The likelihood function has already been computed in Exercise 5.5:

$$L(p) = \prod_{i=1}^n p^{x_i}(1-p)^{1-x_i} = p^{\sum_{i=1}^n x_i}(1-p)^{n-\sum_{i=1}^n x_i}.$$

Taking the natural logarithm turns the product into a sum (because $\log(ab) = \log(a) + \log(b)$):

$$\ell(p) = \log\left(p^{\sum_{i=1}^n x_i}(1-p)^{n-\sum_{i=1}^n x_i}\right)$$
$$= \log\left(p^{\sum_{i=1}^n x_i}\right) + \log\left((1-p)^{n-\sum_{i=1}^n x_i}\right).$$

This expression can be simplified by recalling that $\log(a^b) = b\log(a)$:

$$\ell(p) = \sum_{i=1}^n x_i \log(p) + \left(n - \sum_{i=1}^n x_i\right)\log(1-p).$$

The first-order condition (foc) of the maximization problem $\max_p \ell(p)$ gives (using the chain rule):

$$\frac{\partial \ell(p)}{\partial p} = \frac{1}{p}\sum_{i=1}^n x_i - \frac{1}{1-p}\left(n - \sum_{i=1}^n x_i\right) = 0$$

$$\Longleftrightarrow \frac{\sum_{i=1}^n x_i}{p} = \frac{n - \sum_{i=1}^n x_i}{1-p} \Longleftrightarrow np = \sum_{i=1}^n x_i$$

$$\Longleftrightarrow p = \frac{1}{n}\sum_{i=1}^n x_i = \bar{X}.$$

To check that this is indeed a maximum, note that[4]

$$\frac{\partial^2 \ell(p)}{\partial p^2} = -\frac{\sum_{i=1}^n x_i}{p^2} - \frac{n - \sum_{i=1}^n x_i}{(1-p)^2} < 0.$$

So, $\ell(\cdot)$ is a (strictly) concave function and the unique maximum is attained at $p = \bar{X}$. Therefore, $\hat{p}_{ML} = \bar{X}$.   ◁

A very common mistake is to differentiate the log-likelihood function with respect to x. The log-likelihood function should, however, be differentiated with respect to the parameter θ, because θ is the argument over which the function is maximized.

---

[4]This step is often omitted. For one-dimensional models it is often clear that $\ell(\cdot)$ is concave and unimodal.

Question 6.2. Let $(X_i)_{i=1}^n \overset{iid}{\sim} \mathsf{N}(\mu, \sigma^2)$, where both μ and σ^2 are unknown. Find the ML estimators for μ and σ^2.

Solution 6.2. Again, we first derive the log-likelihood function. Since

$$L(\mu, \sigma^2) = \prod_{i=1}^n \frac{1}{\sqrt{2\pi\sigma^2}} e^{-\frac{1}{2\sigma^2}(x_i-\mu)^2} = (2\pi\sigma^2)^{-\frac{n}{2}} e^{-\frac{1}{2\sigma^2}\sum_{i=1}^n (x_i-\mu)^2},$$

we get (using simlar steps as in Solution 6.1) that

$$\ell(\mu, \sigma^2) = \log\left((2\pi\sigma^2)^{-\frac{n}{2}} e^{-\frac{1}{2\sigma^2}\sum_{i=1}^n (x_i-\mu)^2}\right)$$

$$= -\frac{n}{2}\log(2\pi) - \frac{n}{2}\log(\sigma^2) - \frac{1}{2\sigma^2}\sum_{i=1}^n (x_i - \mu)^2.$$

Differentiating with respect to μ gives

$$\frac{\partial \ell(\cdot)}{\partial \mu} = \frac{1}{\sigma^2}\sum_{i=1}^n (x_i - \mu) = 0 \iff n\mu = \sum_{i=1}^n x_i \iff \mu = \frac{1}{n}\sum_{i=1}^n x_i = \bar{X},$$

whereas differentiation with respect to σ^2 leads to

$$\frac{\partial \ell(\cdot)}{\partial \sigma^2} = -\frac{n}{2\sigma^2} + \frac{1}{2\sigma^4}\sum_{i=1}^n (x_i - \mu)^2 = 0 \iff \sigma^2 = \frac{1}{n}\sum_{i=1}^n (x_i - \mu)^2.$$

Showing that these values actually maximize $\ell(\cdot)$ is a bit tricky and will be omitted here. To conclude,

$$\hat{\mu}_{ML} = \bar{X} \quad \text{and} \quad \hat{\sigma}^2_{ML} = \frac{1}{n}\sum_{i=1}^n (x_i - \bar{X})^2 = \hat{\sigma}^2_X.$$

◁

The next example shows that you have to be careful when maximizing a function: you cannot always blindly take first-order derivatives.

Question 6.3. Let $(X_i)_{i=1}^n \overset{iid}{\sim} \mathsf{U}(0, \theta)$. Find the ML estimator of θ.

Solution 6.3. Convince yourself that (see also Exercise 5.8)

$$L(\theta) = \prod_{i=1}^n \frac{1}{\theta} 1_{(0,\theta)}(x_i) = \frac{1}{\theta^n} 1_{(0,\theta)}(\max x_i),$$

where the last equality follows from the fact that $L(\cdot) \neq 0$ only if all values in $(x_i)_{i=1}^n$ lie in the interval $(0, \theta)$, i.e., only if the largest observation, $\max x_i$, does not exceed θ. Note that this function is not differentiable. When you draw the graph of L (see Figure 6.1), it is obvious that it should hold that $\hat{\theta}_{ML} \geq \max x_i$. In fact, since $\partial L/\partial \theta < 0$ on $[\max x_i, \infty)$, it holds that $\hat{\theta}_{ML} = \max x_i$.

◁

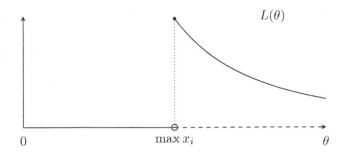

FIGURE 6.1: Likelihood function for a random sample from $U(0, \theta)$.

6.3   Comparing estimators

For a random sample $(X_i)_{i=1}^n \overset{iid}{\sim} U(0, \theta)$, the ML estimator is $\hat{\theta}_{ML} = \max X_i$, i.e., the largest observation. Another possible, and intuitively appealing, estimator would be $\hat{\theta} = 2\bar{X}$, simply because $E(X) = \theta/2$ if $X \sim U(0, \theta)$.[5] In this section we will describe two criteria that are used to distinguish between estimators to determine whether a certain estimator is "better" than another one.

6.3.1   Unbiased estimators

The first, and probably most obvious, question we can ask ourselves is: "if I use the estimator $\hat{\theta}$ (which is a random variable) repeatedly, do I, on average, observe the true value of θ?" If the answer to this question is "yes," then we call the estimator **unbiased**.

Definition 6.5 (unbiased estimator). An estimator $\hat{\theta}$ for θ is *unbiased* if $E(\hat{\theta}) = \theta$.   ◁

Example 6.2. Let $(X_i)_{i=1}^n$ be a random sample from a random variable X for which the expectation μ and the variance σ^2 are finite. Intuitively, you would estimate μ by the sample mean, $\bar{X}$, and σ^2 by the sample variance, $\hat{\sigma}_X^2$. Straightforward computations show that $\bar{X}$ is an unbiased estimator for

---

[5]In fact, in Section 6.4 you will see this is the *moment estimator* of θ.

μ, but that $\hat{\sigma}_X^2$ is not unbiased for σ^2:

$$\mathsf{E}(\bar{X}) = \mathsf{E}\left(\frac{1}{n}\sum_{i=1}^{n} X_i\right) = \frac{1}{n}\sum_{i=1}^{n}\mathsf{E}(X_i) = \frac{1}{n}n\mu = \mu,$$

$$\mathsf{E}(\hat{\sigma}_X^2) = \frac{1}{n}\sum_{i=1}^{n}\mathsf{E}[(X_i - \bar{X})^2] = \mathsf{E}(X_i^2 - 2X_i\bar{X} + \bar{X}^2)$$

$$= \mathsf{Var}(X) + \mathsf{E}(X)^2 - \frac{2}{n}\mathsf{E}(X_iX_1 + \cdots + X_iX_n) + \mathsf{Var}(\bar{X}) + \mathsf{E}(\bar{X})^2$$

$$= \frac{n+1}{n}\mathsf{Var}(X) - \frac{2}{n}[(n-1)\mathsf{E}(X)^2 + \mathsf{E}(X^2)] + 2\mathsf{E}(X)^2$$

$$= \frac{n+1}{n}\mathsf{Var}(X) - \frac{2}{n}[\mathsf{E}(X^2) - \mathsf{E}(X)^2] = \frac{n-1}{n}\sigma^2.$$

However, if we multiply $\hat{\sigma}_X^2$ by $n/(n-1)$, we obtain the **unbiased sample variance**, S_X^2, which *is* an unbiased estimator for σ^2:

$$\frac{n}{n-1}\hat{\sigma}_X^2 = \frac{n}{n-1}\frac{1}{n}\sum_{i=1}^{n}(X_i - \bar{X})^2 = \frac{1}{n-1}\sum_{i=1}^{n}(X_i - \bar{X})^2 = S_X^2, \quad \text{and}$$

$$\mathsf{E}(S_X^2) = \mathsf{E}\left(\frac{n}{n-1}\hat{\sigma}_X^2\right) = \sigma^2.$$

$\triangleleft$

The previous example shows that $\hat{\sigma}_X^2$ is a **biased estimator** for $\mathsf{Var}(X)$. Since $(n-1)/n < 1$, this means that, on average, the estimator $\hat{\sigma}_X^2$ underestimates $\mathsf{Var}(X)$.

Example 6.3. Let $(X_i)_{i=1}^{n} \overset{iid}{\sim} \mathsf{U}(0,\theta)$. Recall that $\hat{\theta}_{ML} = \max X_i$. Also consider the estimator $\hat{\theta} = 2\bar{X}$. Then

$$\mathsf{E}(\hat{\theta}) = \mathsf{E}(2\bar{X}) = 2\mathsf{E}(\bar{X}) = 2\frac{\theta}{2} = \theta,$$

so $\hat{\theta}$ is an unbiased estimator for θ. It can be shown (see Exercise 6.6) that $\hat{\theta}_{ML}$ is not unbiased. Based on this result, you may prefer $\hat{\theta}$ over the maximum likelihood estimator. $\triangleleft$

Example 6.4. Suppose that $(X_i)_{i=1}^{n} \overset{iid}{\sim} \mathsf{Bern}(p)$. Consider the estimators

$$\hat{p}_1 = X_1, \quad \text{and} \quad \hat{p}_2 = \bar{X}.$$

Straightforward computations show that both estimators are unbiased (see also Exercise 6.1). $\triangleleft$

This example shows that unbiasedness in not the be all and end all: we would like a measure for comparing estimators that allows us to argue that $\hat{p}_2$ should be preferred to $\hat{p}_1$. One possibility is to look at the *variance* of unbiased estimators. The idea is that the smaller the variance of an estimator, the smaller the spread of its realizations (the estimates) around its mean (which for an unbiased estimator is the true parameter value). So, an unbiased estimator with a smaller variance is, in that sense, "better." It is easily seen (see Exercise 6.1) that, for the estimators in Example 6.4, it holds that $\text{Var}(\hat{p}_1) > \text{Var}(\hat{p}_2)$.

This leads us to look for the unbiased estimator with the lowest variance, which is called the **best unbiased estimator** (BUE).

Definition 6.6 (best unbiased estimator). Let $(X_i)_{i=1}^n$ be a random sample from a random variable X, the distribution of which depends on an unknown parameter θ. An estimator $\hat{\theta}$ of θ is the *best unbiased estimator* (BUE) if

1. $\hat{\theta}$ is unbiased, and

2. for any other unbiased estimator $\tilde{\theta}$ it holds that $\text{Var}(\hat{\theta}) \leq \text{Var}(\tilde{\theta})$.   ◁

If we restrict attention to *linear* estimators (estimators that are a linear function of the sample), we speak of the **best linear unbiased estimator** (BLUE) as the linear unbiased estimator that has the lowest variance among all linear unbiased estimators.

Theorem 6.1. *Let $(X_i)_{i=1}^n$ be a random sample from a random variable X with mean μ and standard deviation σ. Then the sample mean $\bar{X}$ is the best linear unbiased estimator for μ.*

The proof of this result is actually not very difficult and quite insightful. **Proof.** The claim says something about linear estimators. Any linear estimator can be written as

$$\hat{\mu} = \sum_{i=1}^n a_i X_i,$$

for particular choices of the coefficients $a_1, a_2, \ldots, a_n$. For example, the sample mean is obtained by taking

$$a_1 = a_2 = \cdots = a_n = \frac{1}{n}.$$

We can compute the expectation of any linear estimator:

$$\text{E}(\hat{\mu}) = \text{E}\left[\sum_{i=1}^n a_i X_i\right] = \sum_{i=1}^n \text{E}(a_i X_i) = \sum_{i=1}^n a_i \text{E}(X_i) = \mu \sum_{i=1}^n a_i.$$

Since we are only considering unbiased estimators, this implies that we must choose $(a_1, \ldots, a_n)$ such that

$$\sum_{i=1}^n a_i = 1.$$

For such estimators we can then compute the variance:

$$\mathsf{Var}(\hat{\mu}) = \mathsf{Var}\left[\sum_{i=1}^{n} a_i X_i\right] = \sum_{i=1}^{n} \mathsf{Var}(a_i X_i) = \sum_{i=1}^{n} a_i^2 \mathsf{Var}(X_i) = \sigma^2 \sum_{i=1}^{n} a_i^2,$$

where the second equality holds because all X_i are independent.

For the sample mean we know that $\mathsf{Var}(\bar{X}) = \sigma^2/n$. Now observe that

$$\sum_{i=1}^{n}\left(a_i - \frac{1}{n}\right)^2 = \sum_{i=1}^{n} a_i^2 - \frac{2}{n}\sum_{i=1}^{n} a_i + \sum_{i=1}^{n}\frac{1}{n^2}$$

$$= \sum_{i=1}^{n} a_i^2 - \frac{2}{n} + \frac{1}{n} = \sum_{i=1}^{n} a_i^2 - \frac{1}{n}.$$

Since a sum of quadratic terms is always non-negative, it must, therefore, hold that

$$\sum_{i=1}^{n} a_i^2 \geq \frac{1}{n}.$$

But that means that $\mathsf{Var}(\hat{\mu}) \geq \mathsf{Var}(\bar{X})$ and thus that any linear unbiased estimator for μ always has a variance that is no smaller than the variance of $\bar{X}$. ∎

We now have a good reason for using the sample mean to estimate the population mean: $\bar{X}$ is BLUE. Note that this result holds, regardless of the statistical model. For estimators of other parameters we often do not have such nice results. It turns out, though, that we can find a *lower bound* for the variance of any unbiased estimator.

Theorem 6.2 (Cramér–Rao). *Let* $(X_i)_{i=1}^{n}$ *be a random sample from a random variable* X, *which has a distribution that depends on an unknown parameter* θ. *Suppose that* $\hat{\theta}$ *is an unbiased estimator for* θ. *Under some mild regularity conditions it holds that*

$$\mathsf{Var}(\hat{\theta}) \geq -\left[E\left(\ell''(\theta)\right)\right]^{-1}. \tag{6.3}$$

The right-hand side of (6.3) is called the **Cramér–Rao lower bound** (CRLB). The result might seem baffling at first sight: why would the second derivative of the log-likelihood function have anything to do with the variance of an unbiased estimator? Unfortunately, your mathematics and probability toolbox isn't full enough yet to be able to give a proof. We can, of course, see how it works in an example. The CRLB can often be used to check that an unbiased estimator is BUE: if we can show that the variance of a particular estimator is equal to the CRLB, then no other estimator can have a lower variance; this implies that our estimator must be BUE.

Example 6.5. Let $(X_i)_{i=1}^n \overset{iid}{\sim} \text{Bern}(p)$. From Example 6.1 we know that

$$\ell(p) = \sum_{i=1}^n x_i \log(p) + (n - \sum_{i=1}^n x_i)\log(1-p),$$

$$\frac{\partial \ell(p)}{\partial p} = \frac{\sum_{i=1}^n x_i}{p} - \frac{n - \sum_{i=1}^n x_i}{1-p},$$

and $\hat{p}_{ML} = \bar{X}$.

So,

$$\frac{\partial^2 \ell(p)}{\partial p^2} = -\frac{\sum_{i=1}^n x_i}{p^2} - \frac{n - \sum_{i=1}^n x_i}{(1-p)^2}.$$

We can use this information to directly compute the Cramér–Rao lower bound.

$$\begin{aligned}
\text{CRLB} &= -\left[\mathsf{E}\left(\frac{\partial^2 \ell(p)}{\partial p^2}\right)\right]^{-1} \\
&= -\left[-\frac{n}{p^2}\mathsf{E}(\bar{X}) - \frac{n}{(1-p)^2} + \frac{n}{(1-p)^2}\mathsf{E}(\bar{X})\right]^{-1} \\
&= -\left[-\frac{n}{p} - \frac{n}{(1-p)^2} + \frac{np}{(1-p)^2}\right]^{-1} \\
&= -\left[-\frac{n(1-p)^2 + np - np^2}{p(1-p)^2}\right]^{-1} \\
&= \frac{1}{n}\frac{p(1-p)^2}{1-p} = \frac{p(1-p)}{n}.
\end{aligned}$$

Note that

$$\mathsf{Var}(\hat{p}_{ML}) = \frac{1}{n}\mathsf{Var}(X) = \frac{p(1-p)}{n} = \text{CRLB},$$

so that $\hat{p}_{ML}$ is BUE for p. ◁

6.3.2 Mean squared error

As we saw, not all estimators are unbiased. That does not mean they are necessarily "bad." Restricting attention only to unbiased estimators may lead us to discard estimators that, even though they are biased, might still be useful.

Definition 6.7 (bias). Let $(X_i)_{i=1}^n$ be a random sample from a random variable X, which has a distribution that depends on an unknown parameter θ. Suppose that $\hat{\theta}$ is an estimator for θ. The *bias* of $\hat{\theta}$ is

$$\text{bias}(\hat{\theta}) = \mathsf{E}(\hat{\theta}) - \theta.$$

◁

So, for unbiased estimators we have $\mathsf{bias}(\hat{\theta}) = 0$.

Example 6.6. The bias of $\hat{\sigma}_X^2$ as an estimator for σ^2 is

$$\mathsf{bias}(\hat{\sigma}_X^2) = \mathsf{E}(\hat{\sigma}_X^2) - \sigma^2 = \frac{n-1}{n}\sigma^2 - \sigma^2 = -\frac{\sigma^2}{n} < 0.$$

This shows that $\hat{\sigma}_X^2$ (on average) underestimates σ^2. ◁

Note that the bias depends on the (unknown) parameter.

Rather than requiring unbiasedness, it might be more useful to look at the *expected error* that an estimator makes. The error can be defined in several ways. It could be taken as the distance of an estimator $\hat{\theta}$ to the true parameter θ, i.e., $|\hat{\theta} - \theta|$. Since $\hat{\theta}$ is a random variable, this distance cannot be computed. Therefore, we could take the mean distance $\mathsf{E}[|\hat{\theta} - \theta|]$. However, as with the definition of the variance, it is often not convenient to use distance, because it is not a differentiable function. So, instead, we take the difference and square it. This leads to the idea of **mean squared error**.

Definition 6.8 (mean squared error). Let $(X_i)_{i=1}^n$ be a random sample from a random variable X, which has a distribution that depends on an unknown parameter θ. Suppose that $\hat{\theta}$ is an estimator for θ. The *mean squared error* (MSE) of $\hat{\theta}$ is

$$\mathsf{MSE}(\hat{\theta}) = \mathsf{E}[(\hat{\theta} - \theta)^2].$$

◁

It turns out that the mean squared error of an estimator admits a decomposition in terms of its variance and its bias.

Theorem 6.3. *Let $(X_i)_{i=1}^n$ be a random sample from a random variable X, which has a distribution that depends on an unknown parameter θ. Suppose that $\hat{\theta}$ is an estimator for θ. Then*

$$\mathsf{MSE}(\hat{\theta}) = \mathsf{Var}(\hat{\theta}) + \mathsf{bias}(\hat{\theta})^2.$$

An obvious way to prove this theorem is to use "brute force" and view it as an exercise in using the rules of expectations and variances.

Proof.

$$\mathsf{MSE}(\hat{\theta}) = \mathsf{E}[(\hat{\theta} - \theta)^2] = \mathsf{E}\left[\left(\left(\hat{\theta} - \mathsf{E}(\hat{\theta})\right) + \left(\mathsf{E}(\hat{\theta}) - \theta\right)\right)^2\right]$$

$$= \mathsf{E}\left[\left(\hat{\theta} - \mathsf{E}(\hat{\theta})\right)^2 + 2\left(\hat{\theta} - \mathsf{E}(\hat{\theta})\right)\left(\mathsf{E}(\hat{\theta}) - \theta\right) + \left(\mathsf{E}(\hat{\theta}) - \theta\right)^2\right]$$

$$\overset{(*)}{=} \mathsf{Var}(\hat{\theta}) + 2\mathsf{E}\left[\hat{\theta}\mathsf{E}(\hat{\theta}) - \hat{\theta}\theta - \mathsf{E}(\hat{\theta})^2 + \theta\mathsf{E}(\hat{\theta})\right] + \left(\mathsf{E}(\hat{\theta}) - \theta\right)^2$$

$$= \mathsf{Var}(\hat{\theta}) + 2\left(\mathsf{E}(\hat{\theta})^2 - \theta\mathsf{E}(\hat{\theta}) - \mathsf{E}(\hat{\theta})^2 + \theta\mathsf{E}(\hat{\theta})\right) + \left(\mathsf{E}(\hat{\theta}) - \theta\right)^2$$

$$= \mathsf{Var}(\hat{\theta}) + \left(\mathsf{E}(\hat{\theta}) - \theta\right)^2,$$

where $(*)$ holds since θ is a constant and, hence, $\mathsf{E}(\theta) = \theta$.   ∎

A more elegant way to prove the result requires a bit more thought, but is less involved.

Proof. First observe that

$$\mathsf{Var}(\hat{\theta} - \theta) = \mathsf{E}(\hat{\theta} - \theta)^2 - [\mathsf{E}(\hat{\theta} - \theta)]^2. \qquad (6.4)$$

Now,

$$\mathsf{Var}(\hat{\theta} - \theta) = \mathsf{Var}(\hat{\theta}),$$

because θ is a constant. For the same reason the second term on the right-hand side of (6.4) can be written as

$$[\mathsf{E}(\hat{\theta} - \theta)]^2 = [\mathsf{E}(\hat{\theta}) - \theta]^2 = \mathsf{bias}(\hat{\theta})^2.$$

The first term on the right-hand side is, by definition, the MSE. Combining all this gives

$$\mathsf{Var}(\hat{\theta}) = \mathsf{MSE}(\hat{\theta}) - \mathsf{bias}(\hat{\theta})^2.$$

Straightforward rewriting gives the result.   ∎

Note that the MSE of unbiased estimators is equal to the variance of the estimator. So, if we choose an estimator in Example 6.4 on the basis of the lower MSE, we would choose $\hat{p}_2$. The following example shows that sometimes a biased estimator may have a lower MSE than an unbiased estimator.

Example 6.7. Let $(X_i)_{i=1}^n \overset{iid}{\sim} \mathsf{U}(0, \theta)$. Recall that $\hat{\theta}_{ML} = \max X_i$ is not an unbiased estimator for θ, but that $\hat{\theta} = 2\bar{X}$ is.

Some basic computations give that

$$\begin{aligned}
\mathsf{MSE}(\hat{\theta}) &= \mathsf{Var}(2\bar{X}) + \mathsf{bias}(\hat{\theta})^2 \\
&= \frac{4}{n}\mathsf{Var}(X) = \frac{4}{n}\frac{\theta^2}{12} \\
&= \frac{\theta^2}{3n}.
\end{aligned}$$

In order to compute the MSE of $\hat{\theta}_{ML}$, we need to know the expectation and variance of $\max X_i$. These are fairly straightforward computations and you are asked to do them in Exercise 6.6. There it is derived that

$$\mathsf{MSE}(\hat{\theta}_{ML}) = \frac{2\theta^2}{(n+2)(n+1)}.$$

Since

$$\frac{2}{(n+2)(n+1)} < \frac{1}{3n},$$

for all $n > 2$ (check this!), we conclude that

$$\mathsf{MSE}(\hat{\theta}_{ML}) < \mathsf{MSE}(\hat{\theta}).$$

Therefore, based on the MSE, you would prefer the ML estimator, even though it is not unbiased.   ◁

The following result shows that estimators should ideally be based on sufficient statistics.[6]

Theorem 6.4 (Rao–Blackwell). *Let $(X_i)_{i=1}^n$ be a random sample from a random variable X, which has a distribution that depends on an unknown parameter θ. Suppose that T is a sufficient statistic for θ and that $\hat{\theta}$ is an estimator for θ. Then there exists an estimator $\check{\theta}$, which depends on the sample $(X_i)_{i=1}^n$ through T only, such that $\mathsf{MSE}(\check{\theta}) \le \mathsf{MSE}(\hat{\theta})$. In fact, $\check{\theta}$ can be chosen such that $\mathsf{bias}(\check{\theta}) = \mathsf{bias}(\hat{\theta})$.*

This theorem shows that, in order to find good estimators, we can ignore those that are not based on a sufficient statistic. For certain statistical models we can say even more.

Definition 6.9 (exponential family). Let X be a random variable with density (mass) f_θ that depends on an unknown (k-dimensional) parameter θ. The distribution of X is in an *exponential family* if f_θ can be written as

$$f_\theta(x) = c(\theta) \exp\left\{ \sum_{j=1}^k \zeta_j(\theta) S_j(x) \right\} h(x),$$

where $h(\cdot)$ and $S_j(\cdot)$ are functions on the range of X, and $c(\cdot)$ and $\zeta_j(\cdot)$ are functions of the parameter θ.   ◁

Note that, by applying the factorization theorem, it is clear that $(S_1, \ldots, S_k)$ is a sufficient statistic for θ.

Example 6.8. Let $X \sim \mathsf{Bern}(p)$. We can write

$$f_p(x) = \mathsf{P}(X = x) = (1 - p) \times \exp\left\{ x \log\left(\frac{p}{1-p} \right) \right\} \times 1.$$

So, the probability functions for a Bernoulli experiment belong to an exponential family with $k = 1$, $c(p) = 1 - p$, $\zeta_1(p) = \log(1/(1-p))$, $T_1(x) = x$, and $h(x) = 1$.   ◁

In exponential families there is a clear link between sufficient statistics and unbiased estimators, as the following result shows.

Theorem 6.5. *Let $(X_i)_{i=1}^n$ be a random sample from a random variable X, which has a distribution that depends on a (k-dimensional) unknown parameter θ. If the density of the (joint) distribution of the random sample is in an exponential family*

$$f_\theta(x_1, \ldots, x_n) = c(\theta) \exp\left\{ \sum_{j=1}^k \zeta_j(\theta) T_j(x) \right\} h(x),$$

---

[6]The remainder of this section can be omitted on a first reading.

then, under some mild assumptions, there exists at most one unbiased estimator for θ that only depends on $(T_1, \ldots, T_k)$.

This gives us the following procedure for finding best unbiased estimators.

1. Show that the density of the random sample belongs to an exponential family.

2. Determine a sufficient statistic T.

3. Find an unbiased estimator $\hat{\theta}$, based on T.

It must then be the case that $\hat{\theta}$ is the best unbiased estimator. After all, according to Theorem 6.5, there exists only one unbiased estimator based on T. This must also be the *best* unbiased estimator, since the Rao–Blackwell theorem tells us that the best unbiased estimator must depend on a sufficient statistic.

Example 6.9. Suppose that $(X_i)_{i=1}^n \overset{iid}{\sim} \mathsf{Exp}(\lambda)$. The (joint) density of the random sample is

$$
\begin{aligned}
f_\lambda(x_1, \ldots, x_n) &= \prod_{i=1}^n \frac{1}{\lambda} e^{-x_i/\lambda} \\
&= \frac{1}{\lambda^n} e^{-\sum_{i=1}^n x_i/\lambda} \\
&= c(\lambda) \exp\{\zeta(\lambda) T(x_1, \ldots, x_n)\} h(x_1, \ldots, x_n),
\end{aligned}
$$

where $c(\lambda) = 1/\lambda^n$, $\zeta(\lambda) = -1/\lambda$, $T(x_1, \ldots, x_n) = \sum_{i=1}^n x_i$, and $h(x_1, \ldots, x_n) = 1$. So, f_λ is in an exponential family. It can easily be found that $\hat{\lambda}_{ML} = \bar{X}$, and that $\hat{\lambda}_{ML}$ is an unbiased estimator for λ. The estimator $\bar{X}$ is, therefore, best unbiased. ◁

6.4 Method of moments

A second general method to find estimators in statistical models is the **method of moments**.[7] The idea behind it is pretty straightforward: base the estimate of a *population* moment by computing the analogous *sample* moment. Following this procedure leads to what we call the **moment estimator**. I identify an estimator obtained in this way by the subscript MM and, since it is an estimator, by a hat.

Even though the basic idea is quite simple, writing down a general definition is quite tricky. Recall from Chapter 2 that we can, in principle, write

---

[7]Feel free to skip this section; it contains material that is not required for the rest of the book.

down as many moments of the distribution of some random variable X as we want (assuming that they exist). In general, these moments will depend on (some of the) unknown parameter(s). Assume that we have k unknown parameters $\theta = (\theta_1, \dots, \theta_k)$, and denote the m-th moment of X by

$$\mu_m(\theta) := \mathsf{E}(X^m).$$

Based on a random sample $(X_i)_{i=1}^n$ from the random variable X, we can compute the corresponding **sample moments**, which depend on the data, but not the parameters. The m-th sample moment is given by

$$M_m(X_1, \dots, X_n) := \frac{1}{n} \sum_{i=1}^n X_i^m.$$

Our job now is to equate k well-chosen population and sample moments so that we can solve for the unknown parameters. This solution will only depend on the data and can thus act as an estimator.

Definition 6.10 (method of moments estimator). Let $(X_i)_{i=1}^n$ be a random sample from a random variable X, which has a distribution that depends on a (k-dimensional) unknown parameter θ. A *method of moments* (MM) *estimator* for θ, denoted by $\hat{\theta}_{MM}$, is obtained by simultaneously solving k equations

$$\mu_{m_1}(\hat{\theta}_{MM}) = M_{m_1}(X_1, \dots, X_n)$$

$$\vdots$$

$$\mu_{m_k}(\hat{\theta}_{MM}) = M_{m_k}(X_1, \dots, X_n),$$

where all m_i, $1, \dots, k$ are distinct. ◁

Usually we take the first k moments, but that does not always work.

Question 6.4. Let $(X_i)_{i=1}^n \overset{iid}{\sim} \mathsf{U}(0, \theta)$. Find the moment estimator for θ.

Solution 6.4. Note that

$$\mathsf{E}(X) = \frac{1}{2}\theta =: \mu_1(\theta) \quad \text{and} \quad M_1(X_1, \dots, X_n) = \bar{X}.$$

Solving the equation $\mu_1(\theta) = M_1(X_1, \dots, X_n)$ gives the MM estimator

$$\hat{\theta}_{MM} = 2\bar{X}.$$

◁

The next example shows a case where more than one parameter has to be estimated.

Question 6.5. Let $(X_i)_{i=1}^n \overset{iid}{\sim} \mathsf{N}(\mu, \sigma^2)$. Find the moment estimators for μ and σ^2.

Solution 6.5. We have that

$$\mathsf{E}(X) = \mu =: \mu_1(\mu, \sigma), \quad M_1(X_1, \dots, X_n) = \bar{X}, \quad \text{and}$$

$$\mathsf{E}(X^2) = \sigma^2 + \mu^2 =: \mu_2(\mu, \sigma), \quad M_2(X_1, \dots, X_n) = \frac{1}{n} \sum_{i=1}^{n} X_i^2.$$

Note that the first moment only depends on μ, whereas the second moment depends on both parameters.

The MM estimator for (μ, σ^2) solves

$$\begin{cases} \hat{\mu}_{MM} = \bar{X} \\ \hat{\sigma}_{MM}^2 + \hat{\mu}_{MM}^2 = \frac{1}{n} \sum_{i=1}^{n} X_i^2 \end{cases}$$

$$\Longleftrightarrow \begin{cases} \hat{\mu}_{MM} = \bar{X} \\ \hat{\sigma}_{MM}^2 = \frac{1}{n} \sum_{i=1}^{n} X_i^2 - \bar{X}^2 \end{cases}$$

$$\Longleftrightarrow \begin{cases} \hat{\mu}_{MM} = \bar{X} \\ \hat{\sigma}_{MM}^2 = \frac{1}{n} \sum_{i=1}^{n} (X_i - \bar{X})^2 \end{cases}$$

$$\Longleftrightarrow \begin{cases} \hat{\mu}_{MM} = \bar{X} \\ \hat{\sigma}_{MM}^2 = \hat{\sigma}_X^2. \end{cases}$$

$\triangleleft$

6.5   A useful asymptotic result

We end this chapter by stating a useful asymptotic result for maximum likelihood estimators, which will make our lives easier in later chapters. It is a CLT-like result, which says that, for large samples, the distribution of an ML estimator is approximately normal. In addition, the ML estimator is **asymptotically efficient**, which means that its variance attains the Cramér–Rao lower bound.

Suppose that you have a random sample $(X_1, \dots, X_n)$ from a random variable X with a distribution that depends on an unknown parameter θ. Suppose that your sample $(X_i)_{i=1}^{n}$ is large. Under some regularity conditions it then holds that

$$\hat{\theta}_{ML} \overset{A}{\sim} \mathsf{N}(\theta, \mathsf{CRLB}).$$

The ML estimator is typically what statisticians call a **consistent estimator**: if the sample gets infinitely large, the probability that $\hat{\theta}_{ML}$ deviates from θ gets infinitesimally small (i.e., $\hat{\theta}_{ML} \overset{P}{\to} \theta$). For the purpose of this book you may always assume that the ML estimator is consistent. This is useful, because, as we saw, the Cramér–Rao lower bound depends, typically, on the unknown parameter θ. A deep result in asymptotic theory (see, for example, van der Vaart (1998) if you'd like a taste of what that looks like) says that, if we replace θ by $\hat{\theta}_{ML}$ in order to obtain an estimator for the Cramér–Rao lower bound, let's denote it by $\widehat{\text{CRLB}}$, then the asymptotic distribution does not change:

$$\hat{\theta}_{ML} \overset{A}{\sim} \mathsf{N}(\theta, \widehat{\text{CRLB}}).$$

Example 6.10. Let $(X_i)_{i=1}^{n} \overset{iid}{\sim} \mathsf{Bern}(p)$, with n large. We have previously derived that $\hat{p}_{ML} = \bar{X}$ and that $\text{CRLB} = p(1-p)/n$. So,

$$\hat{p}_{ML} \overset{A}{\sim} \mathsf{N}\left(p, \frac{p(1-p)}{n}\right).$$

Since we can replace p by the ML estimator, we get

$$\hat{p}_{ML} \overset{A}{\sim} \mathsf{N}\left(p, \frac{\hat{p}_{ML}(1-\hat{p}_{ML})}{n}\right).$$

$\triangleleft$

6.6   Chapter summary

We've learned that a good way in which we can estimate an unknown parameter is the maximum likelihood method. Estimators can be judged in different ways, for example, on the basis of (i) best unbiased estimator, or (ii) the mean squared error. We also stated the asymptotic distribution of the maximum likelihood estimator.

Checklist for a successful statistical analysis

1. Identify the population and the parameter of interest.

2. Identify the random variable, X, that is recorded during the experiment.

3. Choose an appropriate family of distributions for X and link the parameter(s) of interest to the parameter(s) of the family of distributions for X.

4. Find an appropriate estimator for the parameter of interest. Be aware of its properties so that you can defend your choice.

6.7   Exercises and problems

Exercises

Parts of most of these exercises require differentiation and some integration. I indicate those parts with a (∗). If you're not comfortable with calculus, you should still be able to do most other parts.

Exercise 6.1. Suppose that $(X_i)_{i=1}^n \overset{iid}{\sim} \mathsf{Bern}(p)$ and consider the two estimators

$$\hat{p}_1 = X_1, \quad \text{and} \quad \hat{p}_2 = \bar{X}.$$

(a) Show that both estimators are unbiased.

(b) On the basis of the mean squared error, which estimator do you prefer? Explain!

Exercise 6.2. Let $(X_i)_{i=1}^n$ be a random sample from a random variable X with mean μ and standard deviation σ. Two estimators for μ are defined by

$$\hat{\mu} = \bar{X}, \quad \text{and} \quad \check{\mu} = \frac{1}{2}X_1 + \frac{1}{2(n-1)}\sum_{i=2}^n X_i.$$

(a) Find the expected values of $\hat{\mu}$ and $\check{\mu}$.

(b) Find the variances of $\hat{\mu}$ and $\check{\mu}$.

(c) Given your answers to parts (a) and (b), explain which of $\hat{\mu}$ and $\check{\mu}$ you prefer for the purpose of estimating μ.

Exercise 6.3. Suppose that $(X_i)_{i=1}^n \overset{iid}{\sim} \mathsf{Poiss}(\lambda)$.

(a) Write down the likelihood function.

(b) Derive the log-likelihood function.

(c∗) Find the derivative of the log-likelihood function.

(d∗) Show that the maximum likelihood estimator for λ is $\hat{\lambda}_{ML} = \bar{X}$.

(e) Compute the bias of $\hat{\lambda}_{ML}$.

(f) Compute the MSE of $\hat{\lambda}_{ML}$.

(g∗) Show that the Cramér–Rao lower bound for the variance of any estimator for λ is $\mathsf{CRLB} = \lambda/n$.

(h) Derive the asymptotic distribution of $\hat{\lambda}_{ML}$.

Exercise 6.4. Suppose that $(X_i)_{i=1}^n \overset{iid}{\sim} \text{Geo}(p)$.

(a$*$) Show that the maximum likelihood estimator for p is $\hat{p}_{ML} = 1/(1+\bar{X})$.

(b$*$) Show that the Cramér–Rao lower bound for the variance of any estimator for p is $\text{CRLB} = p^2(1-p)/n$.

(c) Derive the asymptotic distribution of $\hat{p}_{ML}$.

Exercise 6.5 ($*$). Consider a random sample $(X_i)_{i=1}^n$ from a random variable X with density

$$f_\theta(x) = \begin{cases} \frac{2}{\theta^2}x & \text{if } 0 \le x \le \theta \\ 0 & \text{otherwise} \end{cases}, \quad \theta > 0.$$

(a) Find the maximum likelihood estimator for θ.

(b) Is this estimator unbiased?

(c) Is there another estimator that has a lower variance?

(d) Find the asymptotic distribution of the estimator found in (a).

Exercise 6.6 ($*$). Suppose that you have a sample $(X_i)_{i=1}^n \overset{iid}{\sim} \text{U}(0,\theta)$, for some $\theta > 0$. We know that $\hat{\theta}_{ML} = \max X_i$. In this exercise we study the random variable $Z = \max X_i$.

(a) For any $z < \theta$, show that $\text{P}(Z < z) = \left(\frac{z}{\theta}\right)^n$.

(b) Derive the density function of Z.

(c) Show that $\text{E}(Z) = \frac{n}{n+1}\theta$.

(d) Show that $\text{E}(Z^2) = \frac{n}{n+2}\theta^2$.

(e) Show that $\text{Var}(Z) = \frac{n}{(n+1)^2(n+2)}\theta^2$.

(f) Show that $\text{MSE}(\hat{\theta}_{ML}) = \frac{2}{(n+1)(n+2)}\theta^2$.

(g) Consider the estimator $\check{\theta} = 2\bar{X}$. Show that $\text{MSE}(\hat{\theta}_{ML}) < \text{MSE}(\check{\theta})$ for all $n > 2$.

Exercise 6.7 ($*$). Let $(X_i)_{i=1}^n$ be a random sample from a random variable X with density
$$f_\theta(x) = \theta x^{\theta-1}, \quad 0 < x < 1,$$
where $\theta > 0$ is an unknown parameter.

(a) Find the maximum likelihood estimator for θ.

(b) Derive the asymptotic distribution of $\hat{\theta}_{ML}$.

(c) Compute $\mathsf{E}[\log(X)]$.

(d) Use Jensen's inequality to show that $\hat{\theta}_{ML}$ is not unbiased.

Exercise 6.8 (*). Suppose that X follows the *Hardy–Weinberg* distribution, so the probability mass function of X is

$$f_\theta(k) = \begin{cases} \theta^2 & \text{if } k = 1 \\ 2\theta(1-\theta) & \text{if } k = 2 \\ (1-\theta)^2 & \text{if } k = 3, \end{cases}$$

where $0 < \theta < 1$. Suppose we observe $X = 1$ two times, $X = 2$ five times, and $X = 3$ three times. Find the ML estimate of θ.

Exercise 6.9 (*). Consider a random sample $(X_i)_{i=1}^n \overset{iid}{\sim} \mathsf{N}(\theta, \theta)$. Compute the maximum likelihood estimator for θ.

Problems

Problem 6.1. Every time a particular investment bank writes an investment brief it picks three stocks as "hot buys." In a random sample of twenty reports it was wrong (i.e., delivered a below market average return) on all "hot buys" eleven times, correct once seven times, and correct twice twice. An investor is interested in the probability that a "hot buy" advice is correct.

   (M) Write down a statistical model for this situation and identify the parameter of interest.

   (A.i) Compute the maximum likelihood estimator and maximum likelihood estimate for this parameter.

   (A.ii) Is the estimator unbiased?

   (A.iii) Compute the MSE of the estimator.

   (A.iv) Derive the asymptotic distribution of the estimator derived in (A.i).

   (D) If the investor requires "hot buy" advice to be correct in at least 30% of cases, should the investor buy the advice of this investment bank?

Problem 6.2. The impact of many medical interventions is measured in *Quality Adjusted Life Years* (QALYs). So, the health benefit of an intervention is measured in the number of years that a patient lives longer because of the intervention, corrected for quality of life. Suppose that, for a particular intervention, the QALYs of seven patients are 10.7, 8.3, 0.6, 5.9, 1.3, 4.6 and 3.7. Assuming that these data are the result of a random sample, estimate the average number of QALYs delivered by the intervention. Carefully motivate your estimate.

Problem 6.3 (∗). Suppose that a fraction μ of pollsters always estimates public opinion correctly (within a reasonable margin of error). All other pollsters get it right in 70% of the cases. Find the maximum likelihood estimate for μ if 80 polls are correct in a sample of 100.

Chapter 7

Confidence Intervals

7.1 Introduction

In the previous chapter we thought about finding appropriate estimators for population parameters. But we also know that there is sampling uncertainty and that, therefore, it is likely that any estimate is not going to be exactly equal to the true parameter value. For example, let $(X_i)_{i=1}^{n}$ be a random sample from a continuous random variable X with distribution $F_\theta(\cdot)$, and let $\hat{\theta}$ be an estimator for θ. Since $\hat{\theta}$ is a function of continuous random variables, it is itself a continuous random variable. We therefore have that

$$\mathsf{P}(\hat{\theta} = \theta) = 0.$$

In other words, we know that our estimate is going to be incorrect *even before we compute it*. So, instead of relying on an estimator that picks a single point, in this chapter we will focus on developing an "interval estimator," called a *confidence interval*. We have already constructed such an interval for the case of estimating the mean based on a large sample in Section 4.2, but here we will encounter many more examples.

7.2 Basic idea

Before introducing the general definition, let's think about what an interval estimator could look like. So, let's consider a random sample $(X_1, \ldots, X_n)$ from a random variable X, which follows a distribution that depends on some unknown parameter, call it θ. If we knew θ, then we could compute an interval (l_α, r_α) around θ, such that, for some given probability $\alpha \in (0,1)$, and some estimator $\hat{\theta}$ for θ, it holds that

$$\mathsf{P}\left(\hat{\theta} \in (l_\theta, r_\theta)\right) = 1 - \alpha.$$

Obviously, this interval will differ for every chosen level α.

Example 7.1. Let's illustrate this with a simple statistical model: $(X_i)_{i=1}^{n} \overset{iid}{\sim}$ $N(\mu, \sigma^2)$, with σ known. Suppose that $\mu = \mu_0$ for some known μ_0. We then know that

$$\bar{X} \sim N(\mu_0, \sigma^2/n), \quad \text{or, equivalently, that} \quad Z := \frac{\bar{X} - \mu_0}{\sigma/\sqrt{n}} \sim N(0, 1).$$

If we take, for example, $\alpha = 0.05$, then we can construct an interval $(l_{0.05}, r_{0.05})$, such that

$$P\left(Z \in (l_{0.05}, r_{0.05})\right) = 0.95.$$

That is, we wish to construct an interval around 0 such that 95% of the probability mass of Z lies inside the interval and 5% lies outside the interval. There are many ways in which such an interval can be constructed. Because $Z \sim N(0, 1)$, and the standard normal distribution is symmetric around 0, it makes most sense to create a symmetric interval around 0. That is, we put a probability mass of 2.5% in the left tail and 2.5% in the right tail.

From probability theory we now know how to construct such an interval using inverse probability calculations. First, take the 0.975 percentile of the standard normal, which we denote by $z_{0.975}$. Hence, the symmetric interval for Z around 0 is $(-1.96, 1.96)$. Then transform this interval into "X-units" to find

$$0.95 = P(-1.96 < Z < 1.96)$$
$$= P\left(\mu_0 - 1.96\frac{\sigma}{\sqrt{n}} < \bar{X} < \mu_0 + 1.96\frac{\sigma}{\sqrt{n}}\right). \tag{7.1}$$

For general α we can use the same procedure to find an interval

$$\mu_0 \pm z_{1-\alpha/2}\frac{\sigma}{\sqrt{n}}, \tag{7.2}$$

around μ_0. Note that this interval does not depend on any random variables and is thus constant. See also Figure 7.1.   ◁

[If this computation went too fast for you, please attempt Exercise 7.2.]

The interval that we have computed should be interpreted as follows: if I draw many samples of size n from the $N(\mu_0, \sigma^2)$ distribution, then in 95% of the cases the realized sample mean $\bar{x}$ will lie in the interval (7.2).

This is well and good, but there is a problem: *we don't know if the parameter μ really takes the value μ_0.* So let's put the reasoning upside down: rather than creating an interval around the statistic based on the known parameter, we construct an interval around the unknown parameter based on the statistic.

Example 7.2 (Example 7.1 cont'd). Let's rewrite Equation (7.1) to construct

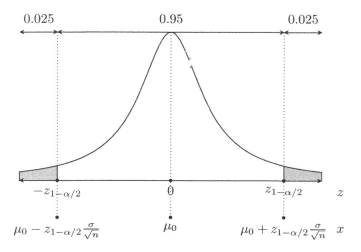

FIGURE 7.1: Symmetric interval around μ_0.

an interval around μ:

$$P\left(\mu - z_{1-\alpha/2}\frac{\sigma}{\sqrt{n}} < \bar{X} < \mu + z_{1-\alpha/2}\frac{\sigma}{\sqrt{n}}\right) = 1 - \alpha$$

$$\Longleftrightarrow P\left(\bar{X} - z_{1-\alpha/2}\frac{\sigma}{\sqrt{n}} < \mu < \bar{X} + z_{1-\alpha/2}\frac{\sigma}{\sqrt{n}}\right) = 1 - \alpha. \qquad (7.3)$$

We now have an interval around μ based on $\bar{X}$:

$$\bar{X} \pm z_{1-\alpha/2}\frac{\sigma}{\sqrt{n}}. \qquad (7.4)$$

This interval does not depend on any unknown quantity. However, since its endpoints are statistics and thus unknown a priori, we call it a **random interval**. ◁

An interval of the form (7.4) is called a **confidence interval**. It gives an interval estimator (L_α, R_α) for the unknown parameter θ, based on some estimator $\hat{\theta}$. Since the interval depends on random variables, its *realization* is different for every sample. Once you've constructed a confidence interval and *before you look at the data*, you can say that "the procedure that I use guarantees that the true parameter θ lies in the (random) interval (L_α, R_α) in a fraction $1 - \alpha$ of all the possible samples that I can draw." Once you have *computed* the confidence interval for the particular sample that you have, you cannot talk about probabilities anymore. At that point you just have some numbers and the true parameter either lies in the interval or not.[1] Hence, a

---

[1]Think about the soccer player and the penalty again.

statistician says something like: "I am 95% confident that the true parameter lies in the computed interval." Note that, in the "two worlds" analogy, the probability statement in (7.3) is about the sample world, whereas the confidence statement is about the real world.

Example 7.3 (Example 7.2 cont'd). Suppose that $\sigma = 5$ and that $n = 25$. If we observe $\bar{x} = 2$, then we can compute the sample value of the 95% confidence interval:

$$\bar{x} \pm z_{1-\alpha/2} \frac{\sigma}{\sqrt{n}} = 2 \pm 1.96 \frac{5}{5} = (0.04, 3.96).$$

So, we can now say that we are 95% confident that μ lies between 0.04 and 3.96.   ◁

All this leads to the following general definition.

Definition 7.1 (confidence interval). Consider a random sample $(X_i)_{i=1}^n$ from a random variable X which has a distribution that depends on an unknown parameter θ. A *confidence interval* of level $1 - \alpha \in (0,1)$ is a (random) interval (L, R), where L and R are statistics with the property that

$$\mathsf{P}(L \leq \theta \leq R) = 1 - \alpha.$$

◁

Note that

1. L and R are random variables, while θ is a constant (unknown) parameter.

2. since L and R are random variables they cannot depend on unknown parameters.

This is all there is to it really. The remainder of this chapter contains several important examples that illustrate the concept of a confidence interval.[2]

7.3   Confidence intervals for means

7.3.1   Mean of a normal population with variance known

Let $(X_i)_{i=1}^n \overset{iid}{\sim} \mathsf{N}(\mu, \sigma^2)$, where σ is known. This is exactly the case that we discussed in Examples 7.1–7.3. There we derived that a $1 - \alpha$ confidence interval for μ is

$$\bar{X} \pm z_{1-\alpha/2} \frac{\sigma}{\sqrt{n}}. \tag{7.5}$$

---

[2]Study tip: do not try to memorize every single example. Rather, try to really understand the main ideas. After careful study, try to replicate all the examples yourself *without* *referring to the text*.

7.3.2 Mean of a normal population with variance unknown

In many cases we will not know the true value of the standard deviation σ. So, let $(X_i)_{i=1}^n \overset{iid}{\sim} \mathsf{N}(\mu, \sigma^2)$, with both μ and σ unknown. As mentioned in Chapter 4, we call σ a *nuisance parameter*, because we are only interested in it in order to obtain an estimator for the mean μ. In the previous case we based our interval on the random variable

$$Z = \frac{\bar{X} - \mu}{\sigma/\sqrt{n}}. \tag{7.6}$$

Here we cannot do that because we don't know σ. To see this, suppose that we were to proceed in the same way as before. We would then derive the interval

$$\bar{X} \pm z_{1-\alpha/2}\frac{\sigma}{\sqrt{n}},$$

which we could not actually compute from the data.

Ideally, we would replace σ in (7.6) by an *estimator*, say the unbiased estimator S_X, which gives the random variable

$$T := \frac{\bar{X} - \mu}{S_X/\sqrt{n}}.$$

The question now is: what is the distribution of T? Can it be $\mathsf{N}(0,1)$? We have replaced the *parameter* σ by the *estimator* S_X. By doing so we have introduced more uncertainty. The distribution of T should, therefore, have a wider spread than the distribution of Z. William Gosset, who worked for the Guinness Brewery in Dublin, answered the question of how much wider: T follows a t-distribution with $n-1$ degrees of freedom.[3]

Theorem 7.1 (Guinness result). *Suppose that* $(X_i)_{i=1}^n \overset{iid}{\sim} \mathsf{N}(\mu, \sigma^2)$, *with* σ *unknown. Then*

$$T = \frac{\bar{X} - \mu}{S_X/\sqrt{n}} \sim t_{n-1}, \tag{7.7}$$

where S_X^2 *is the unbiased sample variance.*

You are asked to prove this result in Exercise 7.3.

The shape of the density of the t-distribution is very similar to the standard normal, but with slightly fatter tails; see Figure 7.2. The larger the sample size n, the more the t-distribution looks like the standard normal. This makes intuitive sense: if we have a larger sample, then there is less sampling uncertainty and S_X is a more precise estimator for σ.

---

[3]Gosset (1876–1937) was concerned with quality control of the stout being produced by Guinness. As he was subject to corporate confidentiality, he could not publish under his own name, but used the nom de plume "Student." In his honour, the t-distribution is often referred to as "Student's t-distribution." Gosset's work shows the importance of transferable skills: he was asked by his employer to solve a particular problem and, given that he couldn't find the solution in any textbook of the time, used his statistical knowledge to come up with a new answer.

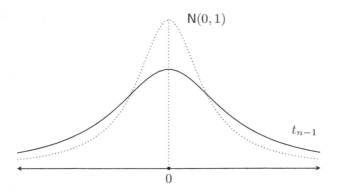

FIGURE 7.2: The standard normal and t_{n-1} distributions.

Now we are ready to find a $1 - \alpha$ confidence interval for μ. Note that, just like the standard normal, the t-distribution is symmetric around 0. As before we shall construct a symmetric interval around 0. Denote the $1-\alpha/2$-percentile of the t_{n-1}-distribution by $t_{n-1;1-\alpha/2}$, i.e.,[4]

$$P(-t_{n-1;1-\alpha/2} \leq T \leq t_{n-1;1-\alpha/2}) = 1 - \alpha.$$

We then find that

$$
\begin{aligned}
1 - \alpha &= P(-t_{n-1;1-\alpha/2} \leq T \leq t_{n-1;1-\alpha/2}) \\
&= P\left(-t_{n-1;1-\alpha/2} < \sqrt{n}\frac{\bar{X} - \mu}{S_X} < t_{n-1;1-\alpha/2}\right) \\
&= P\left(\bar{X} - t_{n-1;1-\alpha/2}\frac{S_X}{\sqrt{n}} < \mu < \bar{X} + t_{n-1;1-\alpha/2}\frac{S_X}{\sqrt{n}}\right).
\end{aligned}
$$

Therefore,

$$\bar{X} \pm t_{n-1;1-\alpha/2}\frac{S_X}{\sqrt{n}},$$

is a $1 - \alpha$ confidence interval for μ. Compare this with (7.5) and note the differences and similarities.

7.3.3 Mean of an unknown distribution based on a large sample

Let $(X_i)_{i=1}^n$ be a random sample from a random variable X with $\mathsf{E}(X) = \mu$ and $\mathsf{Var}(X) = \sigma^2$. Since the distribution of X is not specified we cannot derive the sampling distribution of $\bar{X}$. However, we can find an *approximate* $1 - \alpha$

---

[4]These percentiles can be found, for example, in Table 3.1 of Neave (1978). Note that the last line of this table gives the percentiles of the standard normal distribution.

confidence interval for μ, using the central limit theorem (CLT). In fact, we already did this in Section 4.2. Let's briefly repeat that construction here. Define

$$Z := \frac{\bar{X} - \mu}{\hat{\sigma}_X / \sqrt{n}},$$

where $\hat{\sigma}_X$ is the sample standard deviation. From the CLT we know that

$$Z \overset{A}{\sim} \mathsf{N}(0, 1).$$

So, asymptotically it holds that

$$1 - \alpha \approx \mathsf{P}(|Z| < z_{1-\alpha/2}) = \mathsf{P}\left(\bar{X} - z_{1-\alpha/2}\frac{\hat{\sigma}_X}{\sqrt{n}} < \mu < \bar{X} + z_{1-\alpha/2}\frac{\hat{\sigma}_X}{\sqrt{n}}\right).$$

This then gives the approximate $1 - \alpha$ confidence interval:

$$\bar{X} \pm z_{1-\alpha/2}\frac{\hat{\sigma}_X}{\sqrt{n}}.$$

7.4 Confidence interval for any parameter based on a large sample

Suppose that you have a random sample $(X_i)_{i=1}^{n}$ from a random variable X which has a distribution that depends on an unknown parameter θ. Recall that the maximum likelihood estimator $\hat{\theta}_{ML}$ has the following asymptotic distribution (under some mild regularity conditions):

$$\hat{\theta}_{ML} \overset{A}{\sim} \mathsf{N}(\theta, \widehat{\mathrm{CRLB}}),$$

where $\widehat{\mathrm{CRLB}}$ is an estimator for the Cramér–Rao lower bound.

Using exactly the same reasoning as before, we can derive an approximate $1 - \alpha$ confidence interval for θ (see Exercise 7.6):

$$\hat{\theta}_{ML} \pm z_{1-\alpha/2}\sqrt{\widehat{\mathrm{CRLB}}}. \tag{7.8}$$

Example 7.4. Let $(X_i)_{i=1}^{n} \overset{iid}{\sim} \mathsf{Bern}(p)$. We wish to find an approximate $1 - \alpha$ confidence interval for p, based on a large sample. Recall that

$$\hat{p}_{ML} = \bar{X}, \quad \text{and} \quad \mathrm{CRLB} = \frac{p(1-p)}{n}.$$

We obtain an estimator for the Cramér–Rao lower bound by replacing the unknown parameter p by the estimator $\hat{p}_{ML}$:

$$\widehat{\mathrm{CRLB}} = \frac{\hat{p}_{ML}(1 - \hat{p}_{ML})}{n}.$$

So, an approximate $1 - \alpha$ confidence interval for p is given by

$$\hat{p}_{ML} \pm z_{1-\alpha/2}\sqrt{\frac{\hat{p}_{ML}(1 - \hat{p}_{ML})}{n}}.$$

An important question is when the CLT gives a good approximation. An often used "rule of thumb" is that both the estimated expected number of "successes" and the estimated expected number of "failures" should exceed 5, i.e., $n\hat{p}_{ML} > 5$ and $n(1 - \hat{p}_{ML}) > 5$.   ◁

Example 7.5. Let $(X_i)_{i=1}^n \overset{iid}{\sim} \mathsf{Geo}(p)$. Recall from Exercise 6.4 that

$$\hat{p}_{ML} = \frac{1}{1 + \bar{X}}, \quad \text{and} \quad \text{CRLB} = \frac{p^2(1 - p)}{n}.$$

We obtain an estimator for the Cramér–Rao lower bound by replacing the unknown parameter p by the estimator $\hat{p}_{ML}$:

$$\widehat{\text{CRLB}} = \frac{\hat{p}_{ML}^2(1 - \hat{p}_{ML})}{n}.$$

So, an approximate $1 - \alpha$ confidence interval for p is given by

$$\hat{p}_{ML} \pm z_{1-\alpha/2}\sqrt{\frac{\hat{p}_{ML}^2(1 - \hat{p}_{ML})}{n}}.$$

◁

7.5 Differences between populations based on large samples

Social scientists often want to make inferences comparing different populations. For example, suppose that a researcher wants to investigate the effectiveness of a new fiscal stimulus for small businesses. The way to do this is to create random samples of businesses that received the stimulus (the "target group") and businesses that did not (the "control group"). You can then try to estimate the difference between, say, the average profits of firms in each group.

7.5.1 Difference between population means

Suppose we have a random sample $(X_{1,i})_{i=1}^{n_1}$ from a random variable X_1 with mean μ_1 and standard deviation σ_1, and a random sample $(X_{2,i})_{i=1}^{n_2}$ from a random variable X_2 with mean μ_2 and standard deviation σ_2. Suppose that we are interested in the mean of the difference, call it μ_D:

$$\mu_D := \mathsf{E}(X_1 - X_2).$$

If n_1 and n_2 are large, then the CLT tells us that

$$\bar{X}_1 \overset{A}{\sim} \mathsf{N}(\mu_1, \sigma_1^2/n_1), \quad \text{and} \quad \bar{X}_2 \overset{A}{\sim} \mathsf{N}(\mu_2, \sigma_2^2/n_2).$$

Define the estimator

$$\hat{\mu}_D := \bar{X}_1 - \bar{X}_2.$$

It is easily established that

$$\mathsf{E}(\hat{\mu}_D) = \mu_D, \quad \text{and} \quad \mathsf{Var}(\hat{\mu}_D) = \frac{\sigma_1^2}{n_1} + \frac{\sigma_2^2}{n_2}.$$

Therefore, it holds that,

$$\hat{\mu}_D \overset{A}{\sim} \mathsf{N}\left(\mu_D, \frac{\sigma_1^2}{n_1} + \frac{\sigma_2^2}{n_2}\right),$$

or, after standardizing,

$$\frac{\hat{\mu}_D - \mu_D}{\sqrt{\frac{\sigma_1^1}{n_1} + \frac{\sigma_2^2}{n_2}}} \overset{A}{\sim} \mathsf{N}(0, 1).$$

We do not know σ_1 and σ_2, but can estimate them by

$$\hat{\sigma}_1^2 = \frac{1}{n_1}\sum_{i=1}^{n_1}(X_{1,i} - \bar{X}_1)^2, \quad \text{and} \quad \hat{\sigma}_2^2 = \frac{1}{n_2}\sum_{i=1}^{n_2}(X_{2,i} - \bar{X}_2)^2.$$

It then follows that

$$Z := \frac{\hat{\mu}_D - \mu_D}{\sqrt{\frac{\hat{\sigma}_1^2}{n_1} + \frac{\hat{\sigma}_2^2}{n_2}}} \overset{A}{\sim} \mathsf{N}(0, 1).$$

Using the standard procedure to derive confidence intervals, we can now compute an approximate $1 - \alpha$ confidence interval:

$$\hat{\mu}_D \pm z_{1-\alpha/2}\sqrt{\frac{\hat{\sigma}_1^2}{n_1} + \frac{\hat{\sigma}_2^2}{n_2}}.$$

7.5.2 Difference between proportions

Suppose that $X_1 \sim \mathsf{Bern}(p_1)$ and $X_2 \sim \mathsf{Bern}(p_2)$, with X_1 and X_2 independent. The parameter of interest is $p_D := p_1 - p_2$. It stands to reason to use the estimator $\hat{p}_D := \hat{p}_1 - \hat{p}_2$. Since

$$\hat{p}_1 \overset{A}{\sim} \mathsf{N}\left(p_1, \frac{p_1(1 - p_1)}{n_1}\right) \quad \text{and} \quad \hat{p}_2 \overset{A}{\sim} \mathsf{N}\left(p_2, \frac{p_2(1 - p_2)}{n_2}\right),$$

it holds that

$$\hat{p}_D \overset{A}{\sim} \mathsf{N}\left(p_D, \frac{p_1(1-p_1)}{n_1} + \frac{p_2(1-p_2)}{n_2}\right),$$

or, after standardizing,

$$\frac{\hat{p}_D - p_D}{\sqrt{\frac{p_1(1-p_1)}{n_1} + \frac{p_2(1-p_2)}{n_2}}} \overset{A}{\sim} \mathsf{N}(0,1).$$

Estimating p_1 and p_2 by the sample proportions $\hat{p}_1$ and $\hat{p}_2$, respectively, it then follows that

$$Z := \frac{\hat{p}_D - p_D}{\sqrt{\frac{\hat{p}_1(1-\hat{p}_1)}{n_1} + \frac{\hat{p}_2(1-\hat{p}_2)}{n_2}}} \overset{A}{\sim} \mathsf{N}(0,1).$$

Using the standard procedure we can now compute an approximate $1 - \alpha$ confidence interval:

$$\hat{p}_D \pm z_{1-\alpha/2}\sqrt{\frac{\hat{p}_1(1-\hat{p}_1)}{n_1} + \frac{\hat{p}_2(1-\hat{p}_2)}{n_2}}.$$

7.6 Confidence intervals for the variance of a normal sample

So far, all confidence intervals have been of the form

$$\text{estimator for parameter} \pm c\sqrt{\text{variance of estimator}},$$

where c is some constant that depends on the confidence level $1 - \alpha$ and the sample size (for example, $z_{1-\alpha/2}$ or $t_{n-1;1-\alpha/2}$). That is, they have all been symmetric intervals around a point estimator. We show in this section that not all confidence intervals are necessarily symmetric.

Let $(X_i)_{i=1}^n \overset{iid}{\sim} \mathsf{N}(\mu, \sigma^2)$, with μ and σ^2 unknown. We want to find a confidence interval for σ^2. Using Appendix C.1 we find that

$$C := (n-1)\frac{S_X^2}{\sigma^2} \sim \chi^2_{n-1}.$$

Consider Examples 7.1–7.3. The probability mass inside the confidence interval is $1 - \alpha$, so the mass outside it is α. In assigning this mass we have chosen to assign a probability mass $\alpha/2$ to each of the left tail and the right tail of the distribution. See the left panel of Figure 7.3. Because $\mathsf{N}(0,1)$ is a symmetric distribution, we can find a value for k such that $\mathsf{P}(|Z| < k) = 1-\alpha$. The χ^2 distribution, however, is not symmetric. So, we cannot find just one

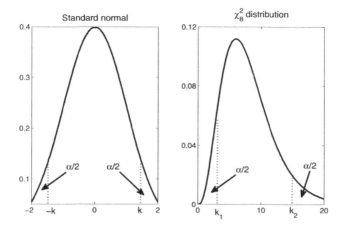

FIGURE 7.3: The standard normal (left panel) and χ_8^2 (right panel) density functions.

value for k. If we decide to retain the policy of assigning $\alpha/2$ of the probability mass to the left tail and $\alpha/2$ to the right tail, then we have to find two values, k_1 and k_2, such that

$$P(C < k_1) = \frac{\alpha}{2}, \quad \text{and} \quad P(C > k_2) = 1 - P(C \le k_2) = \frac{\alpha}{2}, \quad \text{i.e.,}$$

$$k_1 = \chi_{n-1,\alpha/2}^2, \quad \text{and} \quad k_2 = \chi_{n-1,1-\alpha/2}^2.$$

Here $\chi_{n-1,p}^2$ denotes the p-th percentile of the χ_{n-1}^2 distribution. See the right panel of Figure 7.3.

With k_1 and k_2 in hand, we then get

$$1 - \alpha = P(k_1 < C < k_2) = P\left(k_1 < (n-1)\frac{S_X^2}{\sigma^2} < k_2\right)$$

$$= P\left(\frac{k_1}{(n-1)S_X^2} < \frac{1}{\sigma^2} < \frac{k_2}{(n-1)S_X^2}\right)$$

$$= P\left((n-1)\frac{S_X^2}{k_2} < \sigma^2 < (n-1)\frac{S_X^2}{k_1}\right),$$

so that

$$\left((n-1)\frac{S_X^2}{k_2}, (n-1)\frac{S_X^2}{k_1}\right),$$

is a $1 - \alpha$ confidence interval for σ^2.

7.7   Chapter summary

We've learned that it is possible to derive an interval estimator for an unknown parameter of interest. In general, this is done by deriving the (asymptotic) distribution of a carefully chosen statistic, usually based on some point estimator. One then proceeds by finding appropriate boundaries by putting a probability mass $\alpha/2$ in the left tail and $\alpha/2$ in the right tail of this distribution. Rewriting gives an interval around the parameter, based on the point estimator.

A computed $1 - \alpha$ confidence interval does not imply that the probability that the true value of the parameter lies in the interval is $1 - \alpha$. Rather, we have constructed the interval in such a way that, when we apply this procedure over and over again for different samples, the true parameter value lies in a fraction $1 - \alpha$ of the thus computed intervals.

Checklist for a successful statistical analysis

1. Identify the population and the parameter of interest.

2. Identify the random variable, X, that is recorded during the experiment.

3. Choose an appropriate family of distributions for X and link the parameter(s) of interest to the parameter(s) of the family of distributions for X.

4. Find an appropriate estimator for the parameter of interest. Be aware of its properties so that you can defend your choice.

5. Conduct and interpret appropriate statistical inferences, for example:

    (i) confidence intervals for the parameter of interest.

7.8   Exercises and problems

Exercises

Exercise 7.1 (What's wrong?). An exam paper contains the following question.

> A company is interested in the average number of hours per week that 20–30 year olds spend watching YouTube videos. In a sample of 58 20–30 year olds the average is 3.2 hours with a standard deviation of 0.31 hours. Derive and compute a 95% confidence interval.

(a) One student answers as follows.

> Let X denote the average number of hours per week that a randomly chosen 20–30 year old watches YouTube videos. We are told that the sample standard deviation is $\hat{\sigma}_X = 0.31$. In order to obtain the unbiased sample variance, we compute
>
> $$S_X^2 = \frac{n}{n-1}\hat{\sigma}_X^2 = \frac{58}{57}0.31^2 = 0.0978,$$
>
> so that $S_X = 0.313$.
> We know that
>
> $$\frac{\bar{X} - \mu}{S_X/\sqrt{n-1}} \sim t_{n-1},$$
>
> and, from the table, that $t_{57;0.975} \approx t_{60;0.975} = 2.000$. Hence, a 95% confidence interval is given by
>
> $$\bar{x} \pm t_{57;0.975}\frac{S_X}{\sqrt{n-1}} = 3.2 \pm 2.000\frac{0.313}{\sqrt{57}}$$
>
> $$= 3.2 \pm 0.083 = (3.117, 3.283).$$
>
> So, we are 95% confident that the average number of hours per week that a 20–30 year old watches YouTube videos is between 3.117 and 3.283 hours.

What is wrong with this answer?

(b) A second student answers as follows.

$$\bar{X} \pm 1.96\frac{0.31}{\sqrt{n}} = (3.28, 3.12).$$

What is wrong with this answer?

Exercise 7.2. Suppose that $(X_i)_{i=1}^n \overset{iid}{\sim} N(\mu, \sigma^2)$ for some mean μ and variance σ^2.

(a) What is the distribution of the sample mean, $\bar{X}$? Sketch the density function of the distributions of X_1 and $\bar{X}$ in the same figure.

(b) Define the random variable $Z = \sqrt{n}\frac{\bar{X}-\mu}{\sigma}$. What is the distribution of Z? Sketch the density function of this distribution.

(c) In your sketch, fill in a 95% probability mass symmetrically around the mean of Z. How much probability mass is on either side?

(d) Using the tables for the normal distribution, find the precise interval that encompasses 95% of the probability mass in a symmetric interval around the mean of Z.

(e) Find the interval which encompasses 95% of the probability mass of the distribution of $\bar{X}$ in a symmetric interval around μ.

Exercise 7.3 (∗). Suppose that $(X_i)_{i=1}^n \overset{iid}{\sim} N(\mu, \sigma^2)$.

(a) What is the distribution of $\bar{X}$?

(b) Define the random variable $Z = \frac{\bar{X}-\mu}{\sigma/\sqrt{n}}$. What is the distribution of Z?

(c) Show that $\text{Cov}[(X_i - \bar{X}), \bar{X}] = 0$. Argue that this implies that $(X_i - \bar{X})$ and $\bar{X}$ are independent.

(d) Show that S_X^2 and Z are independent, where S_X^2 is the unbiased sample variance.

(e) Define the random variable $Y = \frac{\sum_{i=1}^n (X_i-\bar{X})^2}{\sigma^2}$. What is the distribution of Y? (Hint: Look at Appendix C.)

(f) Define the random variable $T = \frac{\bar{X}-\mu}{S_X/\sqrt{n}}$. Show that $T \sim t_{n-1}$.

Exercise 7.4. Suppose that $(X_{1,i})_{i=1}^{n_1} \overset{iid}{\sim} N(\mu_1, \sigma^2)$, $(X_{2,i})_{i=1}^{n_2} \overset{iid}{\sim} N(\mu_2, \sigma^2)$, that the samples are pairwise independent and that σ is known. [Note that both populations are assumed to have the same standard deviation.]

(a) What are the distributions of $\bar{X}_1$ and $\bar{X}_2$?

(b) What is the distribution of $\bar{X}_1 - \bar{X}_2$?

(c) Derive a $1 - \alpha$ confidence interval for $\mu_D = \mu_1 - \mu_2$.

Exercise 7.5 (∗). Suppose that $(X_{1,i})_{i=1}^{n_1} \overset{iid}{\sim} N(\mu_1, \sigma^2)$, $(X_{2,i})_{i=1}^{n_2} \overset{iid}{\sim} N(\mu_2, \sigma^2)$, that the samples are pairwise independent, and that σ is unknown. Let $\mu_D = \mu_1 - \mu_2$ and $\hat{\mu}_D = \bar{X}_1 - \bar{X}_2$. Define the **pooled variance estimator** S_P^2 by

$$S_P^2 = \frac{(n_1-1)S_{X_1}^2 + (n_2-1)S_{X_2}^2}{n_1 + n_2 - 2}.$$

(a) Use Appendices C.1 and C.2 to show that

$$T := \frac{\hat{\mu}_D - \mu_D}{S_P\sqrt{1/n_1 + 1/n_2}} \sim t_{n_1+n_2-2}.$$

(b) Construct a $1 - \alpha$ confidence interval for μ_D.

Exercise 7.6. Suppose that $(X_i)_{i=1}^n$ is a random sample from a random variable X which has a distribution that depends on an unknown parameter θ. Suppose that n is large. Derive an approximate $1 - \alpha$ confidence interval for θ, based on the maximum likelihood estimator for θ.

Exercise 7.7. Suppose that $(X_i)_{i=1}^n \overset{iid}{\sim} \mathsf{Exp}(\lambda)$. Use the χ^2 distribution to construct a $1 - \alpha$ confidence interval around λ.

Exercise 7.8 (∗). Let $(X_i)_{i=1}^n \overset{iid}{\sim} \mathsf{U}(0, \theta)$, where $\theta > 0$ is unknown. In this exercise we will develop a $1-\alpha$ confidence interval for θ, based on the maximum likelihood estimator

$$\hat{\theta}_{ML} = \max_{i=1,\ldots,n} X_i.$$

First we need to do some probability theory.

(a) Find the distribution function of $\hat{\theta}_{ML}$.

(b) Find an expression for the α-percentile of $\hat{\theta}_{ML}$.

Now we can move to statistics.

(c) Why would it not make sense to construct a symmetric confidence interval around $\hat{\theta}_{ML}$?

(d) Using the result found in (b), construct a $1 - \alpha$ confidence interval for θ.

Problems

Problem 7.1. Recall Problem 5.2. In the remainder, assume that the 169 observations are obtained from a random sample.

(M) Build a statistical model for the situation described in Problem 5.2.

(A) Compute a 95% confidence interval for the fraction who agree.

(D) Discuss the results.

Problem 7.2. You want to rent a furnished one-bedroom apartment in your city next year. The mean monthly rent for a random sample of 10 apartments advertised on a property website is $1,000. Assume that the standard deviation is $92 and that monthly rents are normally distributed. Find a 95% confidence interval for furnished one-bedroom apartments in your city.

Problem 7.3. A researcher is interested in estimating the average number of calories that are contained in pints of Guinness stout. She samples nine pints at random and measures the calorie content (in kcal) of each. The sample mean calorie content is 198.8 kcal and the associated sample standard deviation is 0.60 kcal. Find a 90% confidence interval for the population mean volume.

Problem 7.4. There is concern about the number of cases of misdiagnosis in a particular hospital. A sample of size $n = 150$ reveals three cases of misdiagnosis. Construct a 90% confidence interval for the proportion of misdiagnosed cases. How reliable are your results?

TABLE 7.1: Summary statistics of blood pressure data.

	No. obs.	Mean blood pressure (mmHg)	Sample st. dev. (mmHg)
Group 1	68	88	4.5
Group 2	43	79	4.2

Problem 7.5. A researcher wants to compare the mean blood pressure of two groups of patients. Information from a random sample is summarized in Table 7.1. Compute an approximate 95% confidence interval for the mean difference in blood pressure.

Problem 7.6. Suppose we are interested in whether or not different groups of students are able to solve a particular problem in a statistics course. Random samples of 100 Economics undergraduates and 110 Politics undergraduates are each given the problem to solve. Sixty-five Economics and 45 Politics students solve the problem. Find a 99% confidence interval for the difference in the population proportions.

Problem 7.7. Samples of first-year students are taken at two large universities. The students who have been selected take a test which is marked out of 100%. For the sample of $n_1 = 68$ students selected from the first university, the sample mean and sample variance for the marks are 59 and 29, respectively. For the sample of $n_2 = 54$ students selected from the second university, the sample mean and sample variance for the marks are 61 and 21, respectively. Calculate an approximate 99% confidence interval for the difference between population mean marks.

The next problem is inspired by Brazzale et al. (2007, Section 3.5).

Problem 7.8 (∗). From a 1999 study on recurrent self-harm (Evans et al., 1999) we obtain data on the costs of treatment for two groups of patients with a history of deliberate self-harm. The first group received cognitive behavior therapy and the second group received the conventional therapy. Cost data are often highly skewed, so that an appropriate model seems to be $(X_{1,i})_{i=1}^{n_1} \overset{iid}{\sim} \mathsf{Exp}(\mu_1)$ and $(X_{2,i})_{i=1}^{n_2} \overset{iid}{\sim} \mathsf{Exp}(\mu_2)$. We are interested in the parameter $\lambda := \mu_1/\mu_2$ (i.e., the *ratio* of expected costs). Use Appendix C to find the distributions of

(a) $2\sum_{i=1}^{n_1} X_{1,i}/\mu_1$,

(b) $2\sum_{i=1}^{n_2} X_{2,i}/\mu_2$, and

(c) $Y := \dfrac{\bar{X}_1/\mu_1}{\bar{X}_2/\mu_2}$.

Use the distribution of Y to find a 95% confidence interval for μ_1/μ_2 using the data in Table 7.2.

TABLE 7.2: Costs (in £) of cognitive behavior therapy (group 1) and conventional therapy (group 2).

Group 1	30	172	210	212	335	489	651	1263	1294
	1875	2213	2998	4935					
Group 2	121	172	201	214	228	261	278	279	351
	561	622	694	848	853	1086	1110	1243	2543

Problem 7.9 (*). A manufacturer is concerned about the variability of the levels of impurity contained in consignments of raw material from a supplier. A random sample of 15 consignments showed an (unbiased) standard deviation of 2.36 in the concentration of impurity levels. You may assume that the population follows a normal distribution. Compute a 95% confidence interval for the standard deviation of impurity concentration.

Problem 7.10 (*). You are asked by a consultancy firm to estimate the average number of years that employees in a particular industry stay with the same employer. From previous research you know that the standard deviation is 2.6 years. You are told that you need to be able to estimate the mean with an (absolute) error of at most 0.7 years, with 96% confidence. Write down an appropriate model for this situation, clearly declaring your variables and indicating your assumptions. Also compute the minimum sample size needed to perform the estimation to the required precision.

Problem 7.11 (*). A political party wishes to gain insight in the percentage of votes that it will get at the next general elections. The party asks your consultancy firm to conduct a poll and stipulates that this percentage "should be estimated to an (absolute) precision of 1% with a confidence level of 99%." How many voters need to be polled? How many voters need to be polled if the required confidence level is 95%?

Chapter 8

Hypothesis Testing

8.1 Introduction

In Chapters 6 and 7 we concerned ourselves with estimating unknown parameters. The main goal has been to properly take into account the uncertainty surrounding estimators due to sampling errors. In this chapter we will extend the methodology to answer questions like: "given the sample information, is there evidence that the parameter takes a particular value?" Such questions are of great practical importance. For example, consider a hospital in which concerns are raised about the rate of misdiagnosis. In order to make an informed decision about the hospital's future, we need to know whether the rate of misdiagnosis is higher than the (currently) internationally accepted rate of, say, 1%. Of course, we can look at our estimate, which, say, gives a rate of 2%, and conclude it is higher. However, by now you should realize that this is not an appropriate reasoning, because of sampling uncertainty: another sample might have given you an estimate of, say, 0.8%. So the real question is, if the observed rate is higher than 1%, *taking into account sampling uncertainty*. Statisticians phrase the question in the following way: "is the rate of misdiagnosis in this hospital *statistically significantly* higher than 1%?" In this chapter we will learn how to deal with such questions. We already did this in Section 4.3 for cases where we are interested in the mean of a population and have a large sample available. Here we make the discussion on **hypothesis testing** more general.

8.2 Hypotheses, decisions, and errors

A hypothesis is a statement about the parameters of the model.

Definition 8.1 (statistical hypothesis). Let X be a random variable, the distribution of which depends on an unknown parameter $\theta \in \Theta$. A *statistical hypothesis* H specifies a set Θ', for some subset $\Theta' \subset \Theta$, in which θ is hypothesized to fall. We write $H : \theta \in \Theta'$. ◁

The word "statistical" is usually omitted. We will discuss a framework that is widely used to answer the question: "given two hypotheses and given some data, what do the data tell me about the hypotheses?"

Often there is one particular hypothesis that we are interested in. We call this the **null hypothesis** (or "the null") and denote it by H_0.[1] The null hypothesis is compared to another hypothesis, called the **alternative**, which is denoted by H_1. Hypotheses can be of two types. If the hypothesis fully specifies all the parameters of the model it is called **simple**. Otherwise it is called **composite**.

Definition 8.2 (simple and composite hypotheses). Let X be a random variable, the distribution of which depends on an unknown parameter $\theta \in \Theta$. A hypothesis $H : \theta \in \Theta'$ is *simple* if Θ' is a singleton, i.e., $\Theta' = \{\theta'\}$ for some $\theta' \in \Theta$. The hypothesis is *composite* if Θ' contains at least two values.   ◁

Note that, in the formulation of hypotheses, no reference is made to any data. The random sample will be used to provide evidence for one hypothesis over another.

Example 8.1. You toss a (possibly unfair) coin n times and choose as statistical model $(X_i)_{i=1}^n \overset{iid}{\sim} \mathsf{Bern}(p)$. Based on the random sample, you want to test the null hypothesis $H_0 : p = 1/2$ against the alternative $H_1 : p \neq 1/2$. Here, the null hypothesis is a simple hypothesis, whereas the alternative hypothesis is a composite hypothesis.   ◁

At an abstract level the way we proceed is as follows. We have a random sample $(X_i)_{i=1}^n$ from a random variable X, which has a distribution that depends on an unknown parameter $\theta \in \Theta$. We formulate our hypotheses

$$H_0 : \theta \in \Theta_0, \quad \text{and} \quad H_1 : \theta \in \Theta_1,$$

for some sets $\Theta_0, \Theta_1 \subset \Theta$, such that $\Theta_0 \cap \Theta_1 = \emptyset$.

Question 8.1. Why should it hold that $\Theta_0 \cap \Theta_1 = \emptyset$?

We summarize our data using a statistic $T = T(X_1, \ldots, X_n)$. Our aim is to use this statistic as a measure of evidence and to use its observation in the sample as a basis on which to reject the null hypothesis or not. That is, we are looking for a decision rule d that tells us for each possible realization t of the statistic T whether we reject H_0 or not. Let R be the range of values of T for which H_0 is rejected. Then

$$d(t) = \begin{cases} \text{reject } H_0 & \text{if } T(x_1, \ldots, x_n) \in R \\ \text{do not reject } H_0 & \text{if } T(x_1, \ldots, x_n) \notin R. \end{cases}$$

This procedure is due to Neyman and Pearson, who worked on it during the 1950s, and is therefore called a **Neyman–Pearson test** (NPT).

---

[1]The terminology is due to R.A. Fisher. He considered it the hypothesis "to be nullified." As we will see, the null will be treated in a different way from any other hypothesis in frequentist theory, because it plays Fisher's specific role.

Definition 8.3 (Neyman–Pearson test). Suppose that $(X_i)_{i=1}^n$ is a random sample from a random variable X which has a distribution that depends on an unknown parameter $\theta \in \Theta$. A *Neyman–Pearson test* (NPT) for

$$H_0 : \theta \in \Theta_0 \quad \text{against} \quad H_1 : \theta \in \Theta_1,$$

where $\Theta_0, \Theta_1 \subset \Theta$ and $\Theta_0 \cap \Theta_1 = \emptyset$, consists of a *test statistic* $T = T(X_1, \dots, X_n)$ and a *rejection region* R. The null hypothesis is rejected whenever $t = T(x_1, \dots, x_n) \in R$. ◁

As in Section 4.3, you may wish to think about this in terms of a court case. There are two hypotheses: the defendant is innocent (the null) and the defendant is guilty (the alternative). We gather evidence (our sample) and summarize it for the court (the statistic). The law sets out how this evidence should be interpreted, i.e., whether the hypothesis of innocence is rejected or not (whether the observed statistic lies in the rejection region or not). [At this point you may wish to do Exercise 8.1.]

The particular choice of test statistic T and rejection region R depends, of course, on how we are going to evaluate the decision that we make. Neyman and Pearson did this in terms of an error analysis. In deciding between H_0 and H_1, two types of error can be made. Firstly, you could reject the null hypothesis when it is – in fact – true. This is called a **Type I error**. Secondly, you could accept H_0 when it is – in fact – false. This is called a **Type II error**. Suppose, for example, that we wish to decide between "black" and "white." The decisions and errors that can be made are depicted in Table 8.1. Usually, when we gather empirical evidence we will never be able to unambiguously choose between white and black. Rather, we will observe something on a grey-scale and we will have to decide whether the shade of grey is evidence in favour of white, or whether it is evidence in favour of black.

Definition 8.4 (Type I and Type II errors). Consider an NPT for $H_0 : \theta \in \Theta_0$ against $H_1 : \theta \in \Theta_1$, with statistic T and rejection region R. A *Type I error* occurs if, for a realization t of T, it holds that $t \in R$, while $\theta \in \Theta_0$. A *Type II error* occurs if, for a realization t of T, it holds that $t \notin R$, while $\theta \in \Theta_1$. ◁

TABLE 8.1: Decisions and errors.

		True state of nature	
		White	Black
Decision	White		Type I error
	Black	Type II error	

As with constructing confidence intervals, before we look at the data we can make statements about probabilities, in this case about the probabilities of making Type I and Type II errors. In the "two worlds" analogy, these probability statements are statements about the sample world, not the real world. They are denoted by α and β, respectively, i.e.,

$$\alpha = \mathsf{P}(T \in R | H_0 \text{ is true}) \quad \text{and} \quad \beta = \mathsf{P}(T \notin R | H_1 \text{ is true}). \qquad (8.1)$$

Definition 8.5 (level and power). Consider an NPT with statistic T and rejection region R. The *level* (or size) of the test is the probability of a Type I error, α, and the *power*, Q, is the probability of correctly rejecting an erroneous null hypothesis, i.e., $Q = \mathsf{P}(T \in R | H_1) = 1 - \beta$. ◁

The power of a test is the probability that you don't reject a null that is – in fact – correct, i.e., the probability of not rejecting a null hypothesis that is true. In the court case analogy: the probability that you do not declare an innocent defendant guilty.

Ideally, you would like to keep the probability of both types of errors small. We saw in Section 4.3 that this is not possible: lowering the probability of a Type I error inadvertently increases the probability of a Type II error and thus lowers the power. This can be easily seen in the following, stylized, example.

Example 8.2. Suppose that $X \sim \mathsf{U}(0, \theta)$ and that we want to test the two simple hypotheses

$$H_0 : \theta = 1 \quad \text{against} \quad H_1 : \theta = 2,$$

on the basis of a single observation. So, we want to test if the density of X is given by

$$f_1(x) = 1_{x \in (0,1)}(x), \quad \text{or} \quad f_2(x) = \frac{1}{2} \cdot 1_{x \in (0,2)}(x).$$

These densities are represented in Figure 8.1 by the solid and dotted lines, respectively.

It is obvious that we should reject H_0 if our single observation x is such that $x > 1$, because this observation is impossible under H_0. So, our rejection region R will contain the interval $(1, 2)$, i.e., $R = [c, 2)$, for some $0 < c < 1$. Suppose that we fix a level $\alpha < 1$. That means we have to choose c such that the probability of a Type I error is α, i.e.,

$$\mathsf{P}(X \in R | H_0) = \mathsf{P}(X > c | \theta = 1) = \alpha \iff$$
$$1 - c = \alpha \iff c = 1 - \alpha.$$

This is represented by the light grey area in Figure 8.2. We can now immediately compute the probability of a Type II error:

$$\beta = \mathsf{P}(X < c | \theta = 2) = \frac{1}{2} c = \frac{1}{2}(1 - \alpha).$$

This is represented by the dark grey area in Figure 8.2.

The problem is now immediately clear. If you want to reduce the probability of a Type I error, you need to increase c. But that inadvertently increases

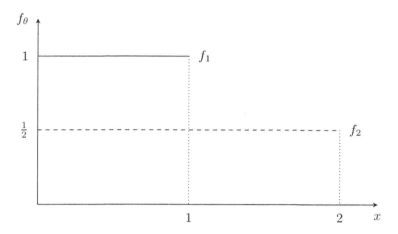

FIGURE 8.1: Densities for Example 8.2.

TABLE 8.2: Critical values and error probabilities for Example 8.2.

c	α	β
0.99	0.01	0.495
0.95	0.05	0.475
0.90	0.10	0.450

the probability of a Type II error. Conversely, if you want to reduce the probability of a Type II error, you need to decrease c. But that inadvertently increases the probability of a Type I error. Table 8.2 gives Type I and II error probabilities for different values of c. ◁

This is a general – and important – phenomenon: for a given sample size, the level of a test can be reduced if, and only if, its power is reduced.

$$\boxed{\text{P(Type I error)} \downarrow \Longleftrightarrow \text{P(Type II error)} \uparrow}$$

We have to decide how we want to balance level and power. Thinking about the court case again, typically we find that convicting an innocent defendant (making a Type I error) is the more serious error. Therefore, we like to control for α (the level) and then find the rejection region R that maximizes the power of the test. If we can find such a region R, we call it the **most powerful test**.

Definition 8.6 (most powerful test). The *most powerful test* of level α of H_0 against H_1 is an NPT with statistic T, rejection region R, level α, and power $1 - \beta$, such that, for any other NPT with statistic $\tilde{T}$, rejection region $\tilde{R}$, and level α, it holds that the power is no larger than $1 - \beta$. ◁

The fact that a test is most powerful does not mean that the probability

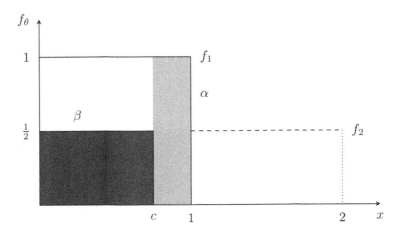

FIGURE 8.2: Probabilities of Type I and Type II errors in Example 8.2

of a Type II error is low; it just means that, among all tests of a certain level, you have found the one with the highest power. Since we usually bias the test against making a Type I error, not being able to reject the null should certainly not lead to *acceptance* of the null. Back to the court case: because we focus on not convicting innocent defendants, failure to produce convincing evidence of guilt does not mean the defendant is innocent. It just means there is not enough evidence to convict. This is why the choice of null hypothesis is important: the Neyman–Pearson set-up is biased *against* rejecting the null. So, it is good statistical practice to take as the null the claim you want to *disprove*.

8.3   What test statistic should we use?

Suppose that we have a sample $(X_i)_{i=1}^n$ from a random variable X which has a distribution that depends on an unknown parameter θ. In this section we will study tests of the type $H_0 : \theta = \theta_0$ against $H_1 : \theta = \theta_1$. In particular, we will be looking for most powerful tests. Neyman and Pearson showed that the choice of test statistic is critical. Once you have the test statistic, the rejection region follows (pretty) straightforwardly. We want to use the sample to see what the relative evidence is for $H_1 : \theta = \theta_1$ as opposed to $H_0 : \theta = \theta_0$. One way to measure the strength of evidence is to look at the likelihood under both hypotheses. So, if the true value of θ is θ_1, then the likelihood of observing our sample is $L(\theta_1)$. If, however, the true value of θ is θ_0, then the likelihood of our sample is $L(\theta_0)$. The *relative* likelihood of θ_1 versus θ_0 is then given by the **likelihood ratio**:

$$\Lambda := \frac{L(\theta_1)}{L(\theta_0)}.$$

If we want to use Λ as our measure of evidence, then we should reject H_0 when the likelihood under θ_1 is large relative to the likelihood under θ_0, i.e., when Λ is large. But how large is "large"? Remember that we are trying to construct a test of a given level α. So we should reject H_0 if $\Lambda \geq c$, where we choose c such that the probability of making a Type I error is α, i.e., where c solves $\mathsf{P}(\Lambda \geq c|\theta = \theta_0) = \alpha$.

Neyman and Pearson showed that this procedure gives the most powerful test. It is summarized in the so-called Neyman–Pearson lemma, the proof of which is too difficult to be included here.

Theorem 8.1 (Neyman–Pearson lemma). *In testing $H_0 : \theta = \theta_0$ against $H_1 : \theta = \theta_1$, the most powerful test of level α has the statistic*

$$\Lambda = \frac{L(\theta_1)}{L(\theta_0)}, \quad \text{and the rejection region} \quad R = [c, \infty),$$

where c is such that $\mathsf{P}(\Lambda \geq c|\theta = \theta_0) = \alpha$.

The number c is called the **critical value** of the test.

For a random sample from a Bernoulli random variable, we can show that the Neyman–Pearson lemma implies that we should base our test on $\sum_{i=1}^n X_i$, or, equivalently, on $\bar{X}$.

Example 8.3. Let $(X_i)_{i=1}^n \overset{iid}{\sim} \text{Bern}(p)$. Suppose we want to test $H_0 : p = p_0$ 🐝 against $H_1 : p = p_1$, with $p_0 < p_1$, using the most powerful test of level α. Note that

$$\Lambda = \frac{p_1^{\sum_{i=1}^n X_i}(1-p_1)^{n-\sum_{i=1}^n X_i}}{p_0^{\sum_{i=1}^n X_i}(1-p_0)^{n-\sum_{i=1}^n X_i}}.$$

The Neyman–Pearson lemma tells us to reject H_0 if $\Lambda \geq c$, where c is such that $\mathsf{P}(\Lambda \geq c|p = p_0) = \alpha$. Unfortunately, it is not easy to find the sampling distribution of Λ, so the number c cannot readily be obtained. If you look carefully at the expression for Λ, however, you will see that it only depends on the data via the (sufficient) statistic $\sum_{i=1}^n X_i$. For that statistic we do know the sampling distribution under H_0: $\sum_{i=1}^n X_i \sim \text{Bin}(n, p_0)$. So let's try and solve the inequality $\Lambda \geq c$ for $\sum_{i=1}^n X_i$. First take logs to get

$$\Lambda \geq c$$
$$\iff \sum_{i=1}^n X_i \log(p_1) + (n - \sum_{i=1}^n X_i) \log(1 - p_1) - \sum_{i=1}^n X_i \log(p_0)$$
$$- (n - \sum_{i=1}^n X_i) \log(1 - p_0)$$
$$\geq \log(c).$$

Now gather all terms which have $\sum_{i=1}^{n} X_i$:

$$\Lambda \geq c$$
$$\iff \sum_{i=1}^{n} X_i \left[\log\left(\frac{p_1}{p_0}\right) - \log\left(\frac{1-p_1}{1-p_0}\right) \right] \geq \log(c) - n \log\left(\frac{1-p_1}{1-p_0}\right).$$

Noting that the term between square brackets is strictly positive (because $p_1 > p_0$), dividing by that term does not alter the direction of the inequality:

$$\Lambda \geq c$$
$$\iff \sum_{i=1}^{n} X_i \geq \log(c) - n \log\left(\frac{1-p_1}{1-p_0}\right) \left[\log\left(\frac{p_1}{p_0}\right) - \log\left(\frac{1-p_1}{1-p_0}\right) \right]^{-1}.$$

The term on the right-hand side does not depend on any unknowns apart from c. So, if we fix c, then we also fix the right-hand side; let's call it d. We now see that finding c such that $P(\Lambda \geq c | p = p_0) = \alpha$ is equivalent to finding some d such that $P(\sum_{i=1}^{n} X_i \geq d | p = p_0) = \alpha$. The latter is possible because we know the sampling distribution of $\sum_{i=1}^{n} X_i$.

For example, if $n = 15$, $p_0 = 1/4$, and $p_1 = 1/2$, then the most powerful test of level $\alpha = 0.05$ is to reject H_0 if $\sum_{i=1}^{n} x_i \geq d$, where

$$P\left(\sum_{i=1}^{n} X_i \geq d \,\Big|\, p = 1/4 \right) = 0.05 \iff d = 8.$$

Therefore, we reject H_0 if we find that $\sum_{i=1}^{n} x_i \geq 8$.

Note that we would find the same d for any alternative $p_1 > p_0$. So, p_1 has no influence on the *level* of the test. It does have an influence on the *power*, since

$$\beta = P\left(\sum_{i=1}^{n} X_i < d \,\Big|\, p = p_1 \right).$$

Under the alternative hypothesis it holds that $\sum_i X_i \sim \mathsf{Bin}(n, p_1)$. Hence, if, for example, $p_1 = 1/2$, then $\beta = P(\sum_{i=1}^{n} X_i \leq 7 | p = 1/2) = 0.5$. Therefore, in testing $H_0 = 1/4$ vs. $H_1 = 1/2$ using the most powerful test of level $\alpha = 0.05$ implies wrongfully not rejecting the null occurs with probability 0.5. By contrast, if $p_1 = 3/4$, it holds that $\beta = 0.0173$ and hence a power of 0.9827. So, even though the probability of a Type I error is the same for both tests, the probability of a Type II error is quite different.   ◁

The following example shows that, when we are dealing with a normally distributed population, we can base the most powerful test on $\bar{X}$.

Example 8.4. Let $(X_i)_{i=1}^{n} \overset{iid}{\sim} \mathsf{N}(\mu, \sigma^2)$, with σ known. Suppose we want to test $H_0 : \mu = \mu_0$ against $H_1 : \mu = \mu_1$, with $\mu_0 < \mu_1$, using the most powerful test at level α. Again we look for the most powerful test by computing the likelihood ratio:

$$\Lambda = \frac{\prod_{i=1}^{n} \frac{1}{\sigma\sqrt{2\pi}} e^{-\frac{1}{2\sigma^2}(X_i-\mu_1)^2}}{\prod_{i=1}^{n} \frac{1}{\sigma\sqrt{2\pi}} e^{-\frac{1}{2\sigma^2}(X_i-\mu_0)^2}} = \frac{e^{-\frac{1}{2\sigma^2}\sum_{i=1}^{n}(X_i-\mu_1)^2}}{e^{-\frac{1}{2\sigma^2}\sum_{i=1}^{n}(X_i-\mu_0)^2}}.$$

From the Neyman–Pearson lemma it follows that the most powerful test is to reject H_0 if $\Lambda \geq c$, where c is such that $P(T \geq c|H_0) = \alpha$. As in the previous example, the sampling distribution of Λ is unknown, so that c cannot be computed.

By taking similar steps as in the previous example, however, we can show that a test based on Λ is equivalent to a test based on $\bar{X}$, for which we do know the sampling distribution. I summarize the steps below:[2]

$$\Lambda \geq c$$

$$\Longleftrightarrow -\frac{1}{2\sigma^2}\sum_{i=1}^{n}(X_i-\mu_1)^2 + \frac{1}{2\sigma^2}\sum_{i=1}^{n}(X_i-\mu_0)^2 \geq \log(c)$$

$$\Longleftrightarrow -\sum_{i=1}^{n}(X_i-\mu_1)^2 + \sum_{i=1}^{n}(X_i-\mu_0)^2 \geq 2\sigma^2\log(c)$$

$$\Longleftrightarrow -\sum_{i=1}^{n}X_i^2 + 2\mu_1\sum_{i=1}^{n}X_i - n\mu_1^2 + \sum_{i=1}^{n}X_i^2 - 2\mu_0\sum_{i=1}^{n}X_i + n\mu_0^2 \geq 2\sigma^2\log(c)$$

$$\Longleftrightarrow (\mu_1-\mu_0)\sum_{i=1}^{n}X_i \geq \sigma^2\log(c) - \frac{n}{2}(\mu_1^2-\mu_0^2)$$

$$\Longleftrightarrow \bar{X} \geq d,$$

for some constant d (which depends on c). So, $P(\Lambda \geq c|\mu = \mu_0) = P(\bar{X} \geq d|\mu = \mu_0)$. Since under the null it holds that

$$\bar{X} \sim N(\mu_0, \sigma^2/n),$$

the most powerful test of level α is to reject H_0 if $\bar{X} \geq d$, where d is such that $P(\bar{X} \geq d|\mu = \mu_0) = \alpha$.

As an example, take $n = \sigma^2 = 10$, $\mu_0 = 0$, and $\mu_1 = 1$. The most powerful test of level $\alpha = 0.05$ can be derived by standardizing $\bar{X}$. Under the null, we know that $\bar{X} \sim N(0, 1)$. So, $P(\bar{X} \geq d|\mu = \mu_0) = 1 - \Phi(d) = 0.05 \Longleftrightarrow d = \Phi^{-1}(0.95) = 1.645$. Therefore, you reject H_0 if $\bar{x} \geq 1.645$. Again, this value does not depend on the alternative value μ_1. Under the alternative, $\bar{X} \sim N(1, 1)$, so that $\beta = P(\bar{X} < 1.645|\mu = 1) = 0.7402$. Thus, the probability of wrongfully accepting the null is roughly 74%. The power of the test is therefore 0.2598. If the alternative had been $H_1 : \mu_1 = 3$, then the probability of a Type II error would have been $\beta = 0.0875$, i.e., a power of 0.9125. ◁

---

[2]It may look complicated but, really, it's no more than high school algebra.

These examples show that the probability of a Type II error can vary greatly for different alternatives, even though the level of the test is the same. This is why you should be very careful with interpreting "not being able to reject the null" as being the same as "accepting the null." Note that the power of a test increases as the alternative gets further away from the null.

Question 8.2. Why do you think that happens?

8.4   Examples of commonly used tests

Let $(X_i)_{i=1}^n$ be a random sample from a random variable X which has a distribution that depends on an unknown parameter θ. We generally distinguish between **one-sided** and **two-sided** tests.

Definition 8.7 (one-sided test). A *one-sided test* is of the form

$$H_0 : \theta = \theta_0, \quad \text{against} \quad H_1 : \theta > \theta_0, \quad \text{or}$$
$$H_0 : \theta = \theta_0, \quad \text{against} \quad H_1 : \theta < \theta_0,$$

for some pre-specified value θ_0.   ◁

Definition 8.8 (two-sided test). A *two-sided test* is of the form

$$H_0 = \theta = \theta_0, \quad \text{against} \quad H_1 : \theta \neq \theta_0,$$

for some pre-specified value θ_0.   ◁

In both one-sided and two-sided tests, we test a simple null against a composite alternative. When we construct NPTs this means that we can always compute the level of the test, because the null fully specifies a unique distribution. We can't compute the power, though, because the alternative specifies not one but a whole set of distributions. This implies that we can't talk about most powerful test anymore. What statisticians do instead is to look for an NPT that is most powerful against every possible value of the alternative. Such a test is called **uniformly most powerful** (UMP). We will not prove the UMP property for any of the tests discussed in this section, but all of the one-sample tests in Sections 8.4.1–8.4.4 are. The tests in Sections 8.4.5 and 8.4.6 are widely used, but are not UMP.

In the remainder of this section, a number of different examples will be discussed. Please look closely at the connection between the reasoning applied here and that used in Chapter 7 to determine confidence intervals.

8.4.1 Test of a normal mean with known variance

Suppose that we have a sample $(X_i)_{i=1}^n \overset{iid}{\sim} \mathsf{N}(\mu, \sigma^2)$, with σ known, and that we wish to conduct a level α one-sided test

$$H_0 : \mu = \mu_0, \quad \text{against} \quad H_1 : \mu > \mu_0. \qquad (8.2)$$

We need to find a region R in which we reject the null hypothesis H_0 and to base this region on some statistic T. In addition, the statement that the level of the test should be α means that the probability of a Type I error should not exceed α, i.e.,

$$\mathsf{P}(T \in R | H_0 \text{ is true}) = \alpha. \qquad (8.3)$$

Recall the confidence interval in Section 7.3.1. We based it on the statistic $Z = (\bar{X} - \mu)/(\sigma/\sqrt{n})$, because we know its distribution: $Z \sim \mathsf{N}(0, 1)$. Unfortunately, Z depends on the unknown parameter μ. However, look at the only probability statement that we are concerned with: (8.3) is *conditional on H_0 being true*. So, *conditional on the null hypothesis being true*

$$Z := \frac{\bar{X} - \mu_0}{\sigma/\sqrt{n}} \sim \mathsf{N}(0, 1). \qquad (8.4)$$

We now have a statistic that we may use for our test. We next need to think about the rejection region R. According to (8.3) we should make sure that the area under the distribution of Z that coincides with R does not exceed α. Let's look at the actual test (8.2) again. If the realization z of Z is large, then it is likely that the true distribution of Z has a mean that is larger than μ_0, i.e., that the alternative hypothesis is true. So, we will reject H_0 if z exceeds some critical value. Since we need to ensure that the probability that we reject if H_0 is true does not exceed α, the area to the right of the critical value can be no more than α. But that means that the critical value is precisely $z_{1-\alpha}$, since, by definition, $\mathsf{P}(Z \leq z_{1-\alpha}) = \Phi(z_{1-\alpha}) = 1 - \alpha$, or, alternatively, that $\mathsf{P}(Z > z_{1-\alpha}) = \alpha$. See also Figure 8.3.

This, then, is our test: reject $H_0 : \mu = \mu_0$ when $z > z_{1-\alpha}$, or, equivalently, when $\bar{x} > c$, where

$$c = \mu_0 + z_{1-\alpha}\frac{\sigma}{\sqrt{n}}.$$

As we saw in Section 8.3, failure to reject H_0 should not lead to acceptance of H_0, because the power of the test can be very low. We can compute the power for different values of the alternative. So, for any $\mu_1 > \mu_0$, the power of the test is

$$Q = \mathsf{P}(\bar{X} \geq c | \mu = \mu_1) = 1 - \mathsf{P}(\bar{X} \leq c | \mu = \mu_1) = 1 - \Phi\left(\frac{c - \mu_1}{\sigma/\sqrt{n}}\right).$$

The two error probabilities are depicted in Figure 8.4.

A test based on the statistic Z is, in general, called a **Z-test**. [At this stage you may wish to do Exercise 8.5.]

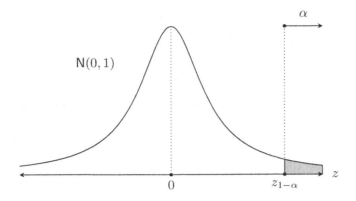

FIGURE 8.3: Right tail of the standard normal distribution.

8.4.2   Test of a normal mean with unknown variance

Suppose that $(X_i)_{i=1}^n \overset{iid}{\sim} \mathsf{N}(\mu, \sigma^2)$, with σ unknown and that we wish to test

$$H_0 : \mu = \mu_0, \quad \text{against} \quad H_1 : \mu < \mu_0, \tag{8.5}$$

at the level α.

Again we need to find a region R in which we reject the null hypothesis H_0. We want to base this region on some statistic T. In addition, the statement that the size of the test should be α means that the probability of a Type I error should not exceed α, i.e.,

$$\mathsf{P}(T \in R | \mu = \mu_0) = \alpha. \tag{8.6}$$

What test statistic can we use? In the previous section we used the Z-statistic, which we based on the confidence interval for a normal mean with variance known. Let's see if we can do that again, but this time, of course, we look at the confidence interval for a normal mean with unknown variance. There we used the fact that

$$\frac{\bar{X} - \mu}{S_X / \sqrt{n}} \sim t_{n-1}.$$

Under the null hypothesis it then holds that

$$T := \sqrt{n} \frac{\bar{X} - \mu_0}{S_X / \sqrt{n}} \sim t_{n-1}, \tag{8.7}$$

which does not depend on any unknown parameter.

We now need to think about the rejection region R. According to (8.3), we should make sure that the area under the distribution of t_{n-1} that coincides

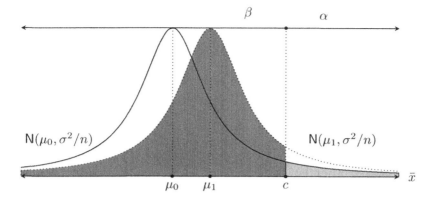

FIGURE 8.4: Error probabilities for a one-sided test of the mean in a normally distributed population.

with R does not exceed α. Following a similar reasoning as before, it makes sense to reject H_0 if the realization of T is *small*. So, we will reject H_0 if t falls short of some critical value. Since we need to ensure that the probability that we reject a correct H_0 does not exceed α, the area to the left of the critical value can be no more than α. That, of course, means that the critical value is $-t_{n-1;1-\alpha}$; see also Figure 8.5.

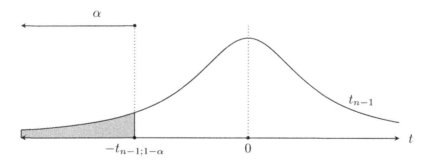

FIGURE 8.5: Left tail of Student's t-distribution.

This, then, is our test: reject $H_0 : \mu = \mu_0$ when $t < -t_{n-1;1-\alpha}$. Equivalently, we can reject H_0 when $\bar{x} < c$, where

$$c = \mu_0 - t_{n-1;1-\alpha} \frac{S_X}{\sqrt{n}}.$$

A test based on the statistic T is, in general, called a **T-test**.

8.4.3   Test of a mean based on a large sample

Let $(X_i)_{i=1}^n$ be a random sample from a random variable X with $\mathsf{E}(X) = \mu$ and $\mathsf{Var}(X) = \sigma^2$. Let's find a test of level α for

$$H_0 : \mu = \mu_0 \quad \text{against} \quad H_1 : \mu \neq \mu_0.$$

In all the tests we discussed before, we found a rejection region in the direction of the alternative. Here the alternative goes in two directions: $\mu > \mu_0$ and $\mu < \mu_0$. That means we can't put the error probability α just to the left or the right. Rather, we put a probability $\alpha/2$ in the left tail and a probability $\alpha/2$ in the right tail. [Note the similarity in reasoning used to construct confidence intervals.]

In order to find an appropriate statistic, recall that the CLT gives that

$$\frac{\bar{X} - \mu}{\hat{\sigma}_X / \sqrt{n}} \overset{A}{\sim} \mathsf{N}(0, 1).$$

Under the null, it therefore holds that

$$Z := \frac{\bar{X} - \mu_0}{\hat{\sigma}_X / \sqrt{n}} \overset{A}{\sim} \mathsf{N}(0, 1).$$

Putting an error probability $\alpha/2$ in the right tail suggests rejecting H_0 in the direction $\mu > \mu_0$ if $z > z_{1-\alpha/2}$. Similarly, putting an error probability $\alpha/2$ in the left tail suggests rejecting H_0 in the direction $\mu < \mu_0$ if $z < z_{\alpha/2} = -z_{1-\alpha/2}$. In shorthand notation: reject H_0 if $|z| > z_{1-\alpha/2}$, or, equivalently, if

$$\bar{x} < \mu_0 - z_{\alpha/2} \frac{\hat{\sigma}_X}{\sqrt{n}}, \quad \text{or} \quad \bar{x} > \mu_0 + z_{1-\alpha/2} \frac{\hat{\sigma}_X}{\sqrt{n}}.$$

See also Figure 8.6.

[This is, of course, exactly the same test we constructed in Section 4.3. At this stage you may also wish to read Section 7.3.3 again.]

Compare this two-sided test to the (approximate) confidence intervals we computed in the previous chapter. You will see that the null hypothesis $H_0 : \mu = \mu_0$ is rejected whenever μ_0 is not in the $1 - \alpha$ confidence interval for μ. That is because we have, essentially, been doing the same thing. For example, we reject $\mu = \mu_0$ (in a large sample) at the 5% level if the realization of $Z = (\bar{X} - \mu_0)/(\hat{\sigma}_X / \sqrt{n})$ exceeds $z_{1-\alpha/2} = 1.96$ in absolute value. But we used the same statistic and critical value to construct a 95% confidence interval for μ.

8.4.4   Test of a proportion based on a large sample

Suppose that $(X_i)_{i=1}^n \overset{iid}{\sim} \mathsf{Bern}(p)$ and that you wish to conduct a test with $H_0 : p = p_0$. If n is large we can use the asymptotic distribution of the ML

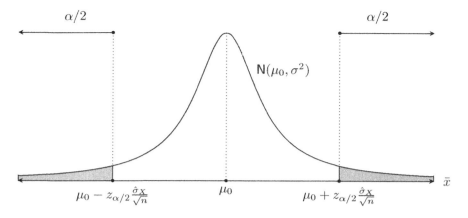

FIGURE 8.6: Two-sided tails of the normal distribution.

estimator:

$$\hat{p}_{ML} \overset{A}{\sim} \mathsf{N}\left(p, \frac{p(1-p)}{n}\right),$$

where $\hat{p}_{ML} = \hat{p}$ is the sample proportion. So, under the null hypothesis, it holds that

$$Z := \frac{\hat{p} - p_0}{\sqrt{p_0(1-p_0)/n}} \overset{A}{\sim} \mathsf{N}(0,1).$$

We can now follow exactly the same procedure as in Section 8.4.3: you reject H_0

1. against $H_1 : p > p_0$ if $z > z_{1-\alpha}$;

2. against $H_1 : p < p_0$ if $z < -z_{1-\alpha}$;

3. against $H_1 : p \neq p_0$ if $|z| > z_{1-\alpha/2}$.

8.4.5 Test of the difference between two means based on a large sample

Suppose that X_1 and X_2 are independent with means μ_1 and μ_2, and variances σ_1^2 and σ_2^2, respectively. Suppose further that you have large random samples of sizes n_1 and n_2, respectively. Define $\mu_D := \mu_1 - \mu_2$ and $\hat{\mu}_D := \bar{X}_1 - \bar{X}_2$. Consider a test with the null hypothesis $H_0 : \mu_D = \mu_0$, for some μ_0.

In Chapter 7 we saw that

$$\frac{\hat{\mu}_D - \mu_D}{\sqrt{\frac{\hat{\sigma}_1^2}{n_1} + \frac{\hat{\sigma}_2^2}{n_2}}} \overset{A}{\sim} N(0,1).$$

So, under the null it holds that

$$Z := \frac{\hat{\mu}_D - \mu_0}{\sqrt{\frac{\hat{\sigma}_1^2}{n_1} + \frac{\hat{\sigma}_2^2}{n_2}}} \overset{A}{\sim} \mathsf{N}(0,1).$$

This is basically the same test as in Section 8.4.3. So we get

1. against $H_1 : \mu_D > \mu_0$ you reject if $z > z_{1-\alpha}$;

2. against $H_1 : \mu_D < \mu_0$ you reject if $z < -z_{1-\alpha}$;

3. against $H_1 : \mu_D \neq \mu_0$ you reject if $|z| > z_{1-\alpha/2}$.

8.4.6 Test of the difference between two proportions in a large sample

Finally, suppose that $X_1 \sim \mathsf{Bern}(p_1)$ and $X_2 \sim \mathsf{Bern}(p_2)$ are independent, and that you have large random samples of sizes n_1 and n_2 of X_1 and X_2, respectively. Define $p_D := p_1 - p_2$ and $\hat{p}_D := \hat{p}_1 - \hat{p}_2$. Suppose that you wish to conduct a test with the null-hypothesis $H_0 : p_D = p_0$. In Section 7.5.2 we saw that

$$\frac{\hat{p}_D - p}{\sqrt{\frac{\hat{p}_1(1-\hat{p}_1)}{n_1} + \frac{\hat{p}_2(1-\hat{p}_2)}{n_2}}} \overset{A}{\sim} N(0,1).$$

So, under the null it holds that

$$Z := \frac{\hat{p}_D - p_0}{\sqrt{\frac{\hat{p}_1(1-\hat{p}_1)}{n_1} + \frac{\hat{p}_2(1-\hat{p}_2)}{n_2}}} \overset{A}{\sim} \mathsf{N}(0,1).$$

We can now again follow the same procedure as in Section 8.4.3: you reject H_0

1. against $H_1 : p_D > p_0$ if $z > z_{1-\alpha}$;

2. against $H_1 : p_D < p_0$ if $z < -z_{1-\alpha}$;

3. against $H_1 : p_D \neq p_0$ if $|z| > z_{1-\alpha/2}$.

[Note that Sections 8.4.4–8.4.6 are pretty much copy-paste from Section 8.4.3. This shows the power of the CLT: as long as you have a large sample you can often (but certainly not always!) base your inferences on some (approximately) normally distributed statistic. Unfortunately, this feature also invites mindless memorization and regurgitation, which can lead to inappropriate use of these tests. Make sure you keep reminding yourself *why* you can use a normal distribution here.]

8.5 *p*-value

Back in Chapter 4 we started out developing our ideas about hypothesis testing by computing the *p*-value. Of course the *p*-value is not just restricted to the case of inference around the mean based on a large sample.

Let $(X_i)_{i=1}^{n}$ be a random sample from a random variable X which has a distribution that depends on an unknown parameter θ.

Definition 8.9 (*p*-value). Consider the test $H_0 : \theta = \theta_0$ against $H_1 : \theta > \theta_0$. For a given statistic T and realization $t = T(x_1, \ldots, x_n)$, the *p-value* is

$$\mathsf{pval} := \mathsf{P}(T(X_1, \ldots, x_n) > T(x_1, \ldots, x_n)|H_0). \qquad (8.8)$$

◁

In words: the *p*-value is the probability that the test statistic is bigger than the *observed* statistic, under the assumption that the null is true. One of the most important 20th century statisticians, R.A. Fisher, came up with this concept, and his idea was that, if H_0 is true and if you were to sample repeatedly, the *p*-value indicates the frequency of samples with a higher value of the statistic. So, if **pval** is very small, then *either* you have observed a rare event, *or* the null hypothesis is not true. Therefore, a low *p*-value is evidence against the null hypothesis. Note that, in the "two worlds" analogy, the *p*-value is a probability statement about the sample world.

Question 8.3. Write down the definition for the *p*-value of a null hypothesis $H_0 = \theta = \theta_0$ against the alternatives $H_1 : \theta < \theta_0$ and $H_1 : \theta \neq \theta_0$. ◁

Example 8.5. Let $(X_i)_{i=1}^{n} \overset{iid}{\sim} \mathsf{N}(\mu, \sigma^2)$, with $\sigma = 3$ known. Consider the null $H_0 : \mu = 1$ against the alternative $H_1 : \mu > 1$. Conditional on the null we know that

$$Z = \frac{\bar{X} - 1}{\sigma/\sqrt{n}} \sim \mathsf{N}(0, 1).$$

If, for example, $n = 50$ and $\bar{x} = 1.8$, we find that $z = 1.88$ and therefore that

$$\mathsf{pval} = \mathsf{P}(Z \geq 1.88|\mu = 1) = 1 - \Phi(1.88) = 1 - 0.9699 = 0.0301.$$

So, either we have observed an event that occurs in 3% of all possible samples (namely, the event $\{Z \geq 1.88\}$) or the hypothesis that $\mu = 1$ is false. This is considered to be strong evidence against the null; see also Figure 8.7.

◁

Many researchers combine the *p*-value and the Neyman–Pearson approach by comparing the *p*-value to some level α ($\alpha/2$ for a two-sided test). For example, if one would have performed a test at level $\alpha = 0.05$, then the

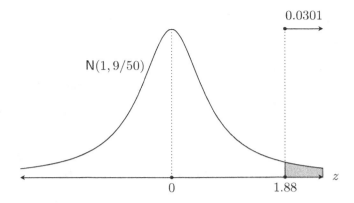

FIGURE 8.7: p-value for a one-sided test.

reasoning is that every p-value smaller than 0.05 would lead to rejection of H_0 in the Neyman–Pearson framework. In the previous example, the observed p-value leads to rejection of the null. For this reason the p-value is sometimes called the **observed level**.

8.6 Statistical significance

In empirical research in the social sciences the term "statistical significance" plays an important part.

Definition 8.10 (statistical significance). A parameter θ is *statistically significantly* larger than/smaller than/different from θ_0 at level α if the null hypothesis $H_0 : \theta = \theta_0$ is rejected against the alternative $H_1 : \theta > \theta_0/H_1 : \theta < \theta_0/H_1 : \theta \neq \theta_0$, respectively, at level α.   ◁

If the null is $H_0 : \theta = 0$ and it is rejected at level α, we usually simply say that "the parameter θ is statistically significant at level α."

Example 8.6. Suppose you are observing a sample $(X_i)_{i=1}^n \overset{iid}{\sim} \mathsf{N}(\mu, \sigma^2)$, with σ known. We have seen in Section 8.4 how we can test the null $H_0 : \mu = 0$ against the alternative $H_1 : \mu \neq 0$: reject H_0 if

$$\left| \frac{\bar{x}}{\sigma/\sqrt{n}} \right| > z_{1-\alpha/2},$$

where $z_{\alpha/2}$ is the $1 - \alpha/2$-percentile of the standard normal distribution. For example, if $\alpha = 0.05$, then $z_{\alpha/2} = 1.96$. Therefore, μ is statistically significant at the 5% level whenever $\left| \frac{\bar{x}}{\sigma/\sqrt{n}} \right| > 1.96$.   ◁

Most researchers interpret statistical significance as evidence that θ is, somehow, important. After all, you are $100(1 - \alpha)\%$ confident that it is not equal to zero. Therefore, the effect measured by θ must actually be there. There is something strange about this reasoning. If your statistical model is formulated as a continuous random variable, you already know that θ will not be *exactly* equal to zero. In fact, you can make any variable statistically significant at any level, as long as you have a large enough sample, because all of the test statistics we have seen in this chapter are increasing in the sample size n. This has led some statisticians to claim that hypothesis tests are nothing more than an elaborate way of measuring the sample size. To see this point, consider the following example.

Example 8.7. Let the statistical model be $(X_i)_{i=1}^{n} \overset{iid}{\sim} \mathsf{N}(\mu, 1)$ and suppose that you observe $\bar{x} = 0.3$. What is the minimum sample size for which μ is statistically significant at the 5% level?

Since we are testing $H_0 : \mu = 0$ against $H_1 : \mu \neq 0$, we know that we reject H_0 if

$$\sqrt{n} \left| \frac{\bar{x}}{\sigma} \right| > z_{1-\alpha/2}.$$

In this case, we reject H_0 if

$$\sqrt{n} \left| \bar{x} \right| > 1.96 \iff \sqrt{n} > \frac{1.96}{0.3} \iff n \geq 43.$$

So, for any sample size exceeding 43 observations, we would conclude that μ is statistically significant. For any smaller sample, we would conclude it is not.

$\triangleleft$

Statistical significance of θ in a test $H_0 : \theta = 0$ against $H_1 : \theta \neq 0$ merely states that your sample is informative enough to recognize, at level α, that θ is not exactly equal to 0. So, only reporting statistical significance is not an appropriate way to convince an audience. If you are an economist, an important additional question is if the result is *economically* significant. For example, suppose that someone tries to explain GDP growth using a number of variables, one of them being the inflation rate. Suppose further that the researcher has a large sample and reports an estimated correlation between inflation and GDP of –0.0001, which is statistically significant at the 1% level. Often, a researcher would say something like "the parameter has the expected sign and is significant" and then stop as if everything that can be said has been said. The researcher is probably trying to tell you that, if inflation goes up 1 percentage point, GDP is expected to go down 0.0001 percentage points. Statistically significant or not, how important is that economically? The power of the test against several alternatives can tell you something about this. See Ziliak and McCloskey (2008) for a (fairly polemic) monograph about the consequences of a narrow focus on the level of a test in several academic fields.

The message of this story is that, no matter how sophisticated the statistical techniques you are employing, you should never stop thinking about what you are actually trying to achieve. By all means, crack a nut with a sledgehammer, but never forget that the ultimate goal is to crack the nut, not to show that you can operate the sledgehammer.

8.7   Chapter summary

We have learned that we can use sampling distributions to test hypotheses. Each test consists of two hypotheses about the (unknown) parameter: a null hypothesis (H_0) and an alternative (H_1). Testing procedures are judged by their level α (probability of a Type I error) and their power $1 - \beta$ (where β is the probability of a Type II error).

We have looked at two broad classes of tests:

1. simple against simple, i.e., $H_0 : \theta = \theta_0$ and $H_1 : \theta = \theta_1$.

2. simple against composite, i.e., $H_0 : \theta = \theta_0$ and $H_1 : \theta \in \Theta_1$, with $\theta_0 \notin \Theta_1$. For these tests we focussed on one-sided and two-sided tests:

     (i) $H_0 : \theta = \theta_0$ and $H_1 : \theta > \theta_0$;

     (ii) $H_0 : \theta = \theta_0$ and $H_1 : \theta < \theta_0$;

     (iii) $H_0 : \theta = \theta_0$ and $H_1 : \theta \neq \theta_0$.

From the Neyman–Pearson lemma we know that the most powerful procedure for testing simple vs. simple hypotheses is of the form: reject the null if an appropriately chosen statistic lies in an appropriately determined rejection region, the shape of which depends on the required level of the test. This has led us to design such tests for one-sided and two-sided tests. The procedure we used was very closely related to the method we used to find confidence intervals in Chapter 7.

A computed test at level α does not imply that the probability of a Type I error is α. Rather, we have constructed the test in such a way that, when we apply this procedure over and over again for different samples, we make a Type I error in a fraction α of the thus computed tests.

Checklist for a successful statistical analysis

1. Identify the population and the parameter of interest.

2. Identify the random variable, X, that iss recorded during the experiment.

3. Choose an appropriate family of distributions for X and link the parameter(s) of interest to the parameter(s) of the family of distributions for X.

4. Find an appropriate estimator for the parameter of interest. Be aware of its properties so that you can defend your choice.

5. Conduct and interpret appropriate statistical inferences, for example:

   (i) confidence intervals for the parameters of interest;

   (ii) hypothesis tests on the parameters of interest.

8.8   Exercises, problems, and discussion

Exercises

Exercise 8.1. A researcher wishes to investigate if the fraction of voters in favour of a new early childhood intervention scheme in a particular country exceeds 50%. The researcher asks a random sample of n voters about their attitude to the proposed scheme.

(M.i) Write down an appropriate statistical model for this situation. What is the parameter of interest?

(M.ii) Formulate an appropriate (simple) null hypothesis and (composite) alternative representing the researcher's objective.

(A.i) What do you think is an appropriate statistic for this model? Why? What is the statistic's distribution?

(A.ii) Construct this test at the 5% level if your sample contains 15 observations.

Exercise 8.2. The claim that a coin is fair is to be tested. Let p denote the probability that "head" is observed when the coin is tossed. The null hypothesis is therefore $H_0 : p = 0.5$. The alternative is $H_1 : p \neq 0.5$. The test is carried out by tossing the coin 10 times and counting the number of "heads" observed. Let X_i denote whether the i-th toss came up heads or tails, so that $(X_i)_{i=1}^{n} \overset{iid}{\sim} \text{Bern}(p)$ is an appropriate statistical model. Two Neyman–Pearson tests are proposed. Both use the statistic $\sum_{i=1}^{n} X_i$, but they have different rejection regions:

- NPT 1: Reject H_0 whenever $\left| \sum_{i=1}^{10} X_i - 5 \right| \geq 4$;

- NPT 2: Reject H_0 whenever $\left| \sum_{i=1}^{10} X_i - 5 \right| \geq 3$.

(A.i) Calculate the Type I error probabilities of both NPTs.

(A.ii) Calculate the Type II error probabilities of both NPTs if $p = 0.4$.

Exercise 8.3. Suppose that you have a sample $(X_i)_{i=1}^n \overset{iid}{\sim} N(\mu, 1)$ of size $n = 9$ and you wish to test the null $H_0 : \mu = 0$ against the alternative $H_1 : \mu = 1$.

(A.i) Construct the most powerful test of level $\alpha = 0.05$.

(A.ii) Draw the distribution of $\bar{X}$ under the different hypotheses and graphically indicate the areas that depict the probabilities of Type I and Type II errors.

(A.iii) Compute the level and power of the test.

(D) You find that $\bar{x} = 0.56$. Draw some conclusions.

Exercise 8.4 (∗). Let $(X_i)_{i=1}^5 \overset{iid}{\sim} \text{Poiss}(\lambda)$. Show that the most powerful test for $H_0 : \lambda = 2$ against $H_1 : \lambda = 4$ at the 5% level rejects H_0 if $\sum x_i \geq 16$. What is the power of this test? (Hint: Follow a similar procedure as in Example 8.3.)

Exercise 8.5. Suppose that $(X_i)_{i=1}^n \overset{iid}{\sim} N(\mu, \sigma^2)$, with σ known. Suppose you want to test the null hypothesis $H_0 : \mu = \mu_0$. Construct an NPT at level α against the alternative

(a) $H_1 : \mu < \mu_0$;

(b) $H_1 : \mu \neq \mu_0$.

Exercise 8.6. Suppose that you have a random sample $(X_i)_{i=1}^n \overset{iid}{\sim} \text{Poiss}(\lambda)$, with n large. Construct an (asymptotic) NPT at level α for the null hypothesis $H_0 : \lambda = \lambda_0$ against the alternative $H_1 : \lambda \neq \lambda_0$.

Exercise 8.7. Suppose that you have a random sample $(X_i)_{i=1}^n \overset{iid}{\sim} \text{Geo}(p)$, with n large. Construct an (asymptotic) NPT at level α for the null hypothesis $H_0 : p = p_0$ against the alternative $H_1 : p > p_0$.

Exercise 8.8. Suppose that $(X_i)_{i=1}^n \overset{iid}{\sim} \text{Exp}(\lambda)$, with λ an unknown parameter.

(a) Find the most powerful test of level α of $H_0 : \lambda = \lambda_0$ against $H_1 : \lambda = \lambda_1$ with $\lambda_1 < \lambda_0$.

Throughout the remainder of this exercise assume that $n = 10$ (which is *not* enough to assume that the CLT can be applied). Find an NPT at level $\alpha = 0.05$ for the test $H_0 : \lambda = \lambda_0$ against the alternative:

(b) $H_1 : \lambda > \lambda_0$;

(c) $H_1 : \lambda < \lambda_0$;

(d) $H_1 : \lambda \neq \lambda_0$.

Problems

Problem 8.1. A new manuscript with eight poems has come to light and scholars wish to establish if the poems are by a particular seventeenth century poet. It is known that works by this poet contain, on average, 9.1 new words (i.e., words that have not been used in the poet's earlier work). Scholarly consensus is that other authors from the same period contain fewer new words than this poet's work. In the newly discovered poems there are, on average, 8.5 new words with a sample standard deviation (unbiased) of 2.85 words.

(M) Develop an appropriate statistical model for this situation. Carefully declare the variable and the parameter(s).

(A) Construct and conduct an appropriate test to investigate the claim that the eight newly found poems are by this particular poet.

(D) Draw some conclusions from your investigation.

Problem 8.2. A newspaper claims that the average waiting time for treatment of a particular disease in a certain hospital is a "shocking, life-threatening" 13 days, whereas government guidelines require patients to be seen within 10 days. The article is based on a random sample of 15 patients who had to wait, on average, 12.62 days.

(M) Build an appropriate statistical model for this situation.

(A) Formulate and develop the most powerful test at the 5% level and compute its power.

(D) Discuss your findings.

Problem 8.3. The average annual income, in $000's, for a certain population is denoted by μ. A random sample of 50 individuals is drawn and data are obtained on annual incomes in $000's. It is found that the sample mean is 22.3 and the sample variance is 12.5. Use this sample information to test $H_0 : \mu = 20$ against $H_1 : \mu > 20$ at the 5% significance level. You should comment on any assumptions that you make.

Problem 8.4. It is claimed that jars of coffee contain, on average, 250 grams of coffee. A customer believes that jars are underweight. She buys eight jars in order to weigh their contents and finds that the sample mean weight is 246.4. The sample variance of weight is $\hat{\sigma}_X^2 = 11.52$. Interpret this evidence and comment on the claim that coffee jars contain, on average, 250 grams of coffee.

Problem 8.5. A large bank with roughly equally sized branches around the country is confronted, on average, with two bad cheques each day. The manager of the branch in your city is suspicious and suspects that the number in her branch could be as high as four bad cheques per day. A random sample over 5 days reveals 14 bad cheques. Advise her on the basis of these results.

Problem 8.6. It is known that, on a certain part of a road, 20% of drivers exceeded the speed limit of 80 km/h by 5 km/h or more. After the police change the pattern of patrols, it is claimed that there has been a substantial reduction in this proportion. Comment on this claim based on evidence from 100 speed measurements among which 10 recorded a speed of at least 80 km/h.

Problem 8.7 (∗)**.** New employees at a company are required to attend a training course. The traditional form of the course lasts 3 days and is based upon lectures and seminars with question-and-answer sessions. A new type of course takes only 2 days and is based upon videos. The new method is preferred if it is not less effective. Effectiveness is assessed by testing employees at the end of their training course. A statistician arranges for 16 randomly selected new employees to attend the traditional course and for 13 randomly selected new employees to attend the new course. After the test scores have been calculated, it is found for the first group, i.e., the 16 employees on the traditional course, that the sample mean score is 48.7 and the sample variance of scores is 20.0. For the second group, which consists of 13 employees, the sample mean score is 47.2 and the sample variance of scores is 15.0. Use these sample values to test the claim that there is no difference between the population mean scores of those attending the traditional and new types of training course. The alternative hypothesis is that the new type of training is less effective than the traditional course. Construct and perform an appropriate test. What do you conclude from the outcome of your test? What assumptions are required to justify the test that you used? [Hint: Use the results from Exercise 7.5.]

Problem 8.8. There is concern about the number of cases of misdiagnosis in a particular hospital. If the occurrence of misdiagnosis is too high, maybe the hospital should be closed, with enormous consequences for the local community. It is therefore paramount that the matter is investigated carefully. By international standards, a fraction of 1% of cases of misdiagnosis is considered as "international best practice." A sample of size $n = 150$ reveals 3 cases of misdiagnosis. Construct and perform a statistical test to investigate the claim that "this hospital has a higher rate of misdiagnosis than is internationally acceptable."

Discussion

This section contains a few exercises that encourage you to think more deeply about some of the inferential procedures that we have encountered.

Discussion 8.1. Health campaigners accuse a particular tobacco manufacturer of advertising with a lower nicotine content than their cigarettes actually possess. The manufacturer, Smoke'n'All, claims its cigarettes contain 1.2 mg of nicotine, on average, per cigarette. The health campaigners, BreathFree, claim it is actually 1.4 mg. Smoke'n'All commissions a statistician to investigate the claim. He gathers the nicotine content of a random sample of n cigarettes and models it as $(X_i)_{i=1}^n \overset{iid}{\sim} N(\mu, \sigma^2)$, where μ is the average nicotine content in cigarettes. He conducts a test $H_0 : \mu = 1.2$ against $H_1 : \mu = 1.4$ and finds no evidence to reject H_0 at the 5% level. You are asked by BreathFree to comment on this research.

(a) Write a short report for BreathFree on the way that the statistician hired by Smoke'n'All did his job.

Suppose that the researcher's sample size was $n = 25$, with a sample mean of $\bar{x} = 1.33$. The standard deviation is known to be $\sigma = 0.44$.

(b) Conduct the test that Smoke'n'All's statistician conducted and verify his conclusion. What is the power of this test?

(c) Conduct the test that Smoke'n'All's statistician *should have* conducted (at the 5% level) and report on your conclusions. What is the power of this test?

BreathFree has accused Smoke'n'All of providing misleading information. Smoke'n'All, in turn, has sued BreathFree for libel. You have been called as an expert witness by the Court and you are asked how it should interpret the evidence. Smoke'n'All has presented evidence based on the test $H_0 : \mu = 1.2$ and $H_1 : \mu = 1.4$, whereas BreathFree has presented the results based on $H_0 : \mu = 1.4$ and $H_1 : \mu = 1.2$.

(d) Comment on the evidence provided by both sides. On the balance of this evidence, which party would you say is correct in its statements?

The following example is taken from Berger (1985, Example 1.8).

Discussion 8.2. You have a sample $(X_i)_{i=1}^n \overset{iid}{\sim} N(\mu, 1)$ and want to test the null $H_0 : \mu = 0$ against the alternative $H_1 : \mu > 0$ at level $\alpha = 0.05$.

(a) Find the rejection region in a sample[3] of size $n = 10^{24}$.

(b) Suppose that the true value of μ is 10^{-10}. What is the power of the test you conducted?

On using the same level in every statistical test of a simple null against a composite hypothesis, Fisher (1973, p. 45) writes:

the calculation is absurdly academic, for in fact no scientific worker has a fixed level of significance at which from year to year, and in all circumstances, he rejects hypotheses; he rather gives his mind to each particular case in the light of his evidence and his ideas.

---

[3]This may look like an unrealistically large sample, but if you think about how much internet traffic is generated every day, this no longer seems so outlandish if your research uses data obtained from internet usage.

(c) Comment on this quote in light of (a) and (b).

Leamer (1978, p. 89) writes:

> we might as well begin by rejecting the null hypothesis [in a one or two-sided NPT] and not sample at all.

(d) Comment on this quote in light of (a) and (b).

Discussion 8.3. Suppose that all authors of a particular journal test their statistical hypotheses at the 5% level.

(a) In the long run, what percentage of articles presents evidence of statistically significant effects that are, in fact, not present?

According to Wikipedia, **publication bias** is "a bias with regard to what is *likely* to be published, among what is *available* to be published" [emphasis added]. It is considered to be an important problem in scientific reporting.[4]

(b) How would publication bias influence the percentage you mentioned under (a)?

Discussion 8.4. You will be given one observation $X \sim N(\mu, 1)$ on a DNA sample that was found at a crime scene. Your client has sent it to a test lab and is awaiting the result. You are asked to test the hypothesis $H_0 : \mu = 0$ against $H_1 : \mu > 0$ at the 5% level.

(a) Construct an appropriate NPT.

A few days later the measurement comes in: $x = 1.6$.

(b) Conduct the NPT constructed in (a).

A week later you talk about this research with a friend, who is a well-known statistician, and you tell her that, after you had conducted your test, you learned that your client could have sent the sample to another lab, in which case you would have gotten an observation $Y \sim N(\mu, 1)$. Your friend now becomes quite agitated and asks you how your client decided between the two labs. After sending your client an email, it transpired that they simply flipped a coin. Your friend, in the meantime, has done some research into the two labs and concludes that, since they use similar technologies, the correlation between measurements is 0.8.

(c) Your friend argues that you should base your inference on the random variable $Z = 0.5X + 0.5Y$. Why?

(d) For this random variable, construct and conduct an appropriate test.

---

[4]For an exposition of the problem in relation to clinical trials, see, for example, Goldacre (2012).

This problem illustrates that frequentist statisticians have to be very careful when designing their statistical model, because they have to be able to compute probabilities over *all* possible observations. In particular, the use of NPTs or p-values implies that your conclusions are drawn based on something that you *could have* observed, but didn't. Jeffreys (1961, Section 7.2) puts it rather nicely:

> What the use of [the p-value] implies, therefore, is that a hypothesis that may be true may be rejected because it has not predicted observable results that have not occurred.

Chapter 9

Linear Regression

9.1 Basic ideas

So far we have introduced several methods that can be used to draw inferences on a single population parameter using a sample. Social scientists are often concerned with how several variables are related to each other. For example, economic theory may say something about how the money supply and inflation are related and an econometrician may want to test that theory on the basis of some data.

In particular, we could postulate that the mean of a certain random variable Y, denoted by μ_Y, could depend on the realization of another random variable X.

Example 9.1. A particular statistics module is assessed by a midterm test and an end-of-term exam. The lecturer thinks that the mean of the score of the end-of-term exam, Y, depends on the score for the midterm, X. After the midterm has taken place, the lecturer can observe the midterm score, x, for every student and use that to model, probabilistically, the score for the end-of-term exam.

A random sample of $n = 10$ students is considered and the midterm scores, x, and the end-of-term exam scores, y, are recorded. The data are given in Table 9.1. A scatter plot of these data can be found in Figure 9.1.   ◁

A possible statistical model could be built by assuming that the distribution of Y, *conditional on* X, is normal with a mean that depends linearly on

TABLE 9.1: Midterm and end-of-term exam scores (%) for a random sample of 10 students.

obs.	midterm	end-of-term	obs.	midterm	end-of term
1	56	55	6	55	55
2	54	62	7	71	64
3	61	53	8	65	67
4	46	59	9	71	75
5	57	75	10	61	54

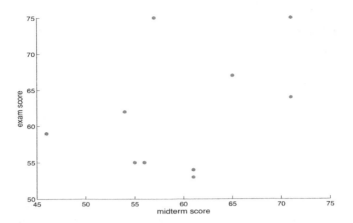

FIGURE 9.1: Scatter plot of midterm and end-of-term exam scores.

the realization of X:

$$Y|\{X = x\} \sim \mathsf{N}(\beta_0 + \beta_1 x, \sigma^2).$$

Note that this is how we use a conditional distribution to model the relation between variables X and Y: the realization of X fixes the distribution of Y. In the context of Example 9.1, each possible realization of the midterm grade fixes a different distribution of the end-of-term grade. This may give you the impression that we consider the value of x to fixed, while we treat Y as random. Many econometricians actually do this, but that leads to some epistemological problems in terms of statistical inference; see Keuzenkamp (1995). The start of (linear) regression analysis is a model of the distribution of some random variable, Y, conditional on another random variable X.

A first question that arises is why the expectation of Y should be a *linear* function of x. Recall from Section 2.5.1 that if

$$X \sim \mathsf{N}(\mu_X, \sigma_X^2), \quad \text{and} \quad Y \sim \mathsf{N}(\mu_Y, \sigma_Y^2),$$

with

$$\sigma_{XY} := \mathsf{Cov}(X, Y) = \mathsf{E}[(X - \mu_X)(Y - \mu_Y)],$$

then, jointly, X and Y have a bivariate normal distribution. Denoting the correlation coefficient by $\rho = \frac{\sigma_{xy}}{\sigma_x \sigma_y}$, we know that, conditional on $\{X = x\}$, Y has the distribution

$$Y|\{X = x\} \sim \mathsf{N}\left(\mu_Y + \rho \frac{\sigma_Y}{\sigma_X}(x - \mu_X), \sigma_Y^2(1 - \rho^2)\right).$$

In particular, the conditional mean of Y given $\{X = x\}$ is

$$\mathsf{E}(Y|X = x) = \beta_0 + \beta_1 x,$$

where

$$\beta_0 = \mu_Y + \rho \frac{\sigma_Y}{\sigma_X} \mu_X \quad \text{and} \quad \beta_1 = \rho \frac{\sigma_Y}{\sigma_X}. \tag{9.1}$$

In this bivariate normal model, the conditional mean of Y is indeed a linear function of X.

This means that, if we want to draw inferences on $\mathsf{E}(Y|X = x)$, then we need to draw inferences on β_0 and β_1. Since $\mathsf{E}(Y|X = x)$ is a linear function of x, we call this statistical model a **linear regression model**.

The assumption of normality is a restrictive one, but the general idea is that we can often model the (conditional) mean of one variable as a linear function of outcomes of other variables. In this book we will only examine the case where there are two variables. In principle, any number of variables can be used (as long as there are enough observations).

Note that the two variables X and Y are treated differently. Specifically, we assume that Y is dependent on X. This is why we call Y the **dependent** (or **endogenous**) variable and X the **independent** (or **exogenous**) variable. Moreover, the random variable X is conditioned upon its realization x.

So, we will be analyzing pairs of data $((x_1, y_1), (x_2, y_2), \ldots, (x_n, y_n))$ which result from a random sample $((X_1, Y_1), \ldots, (X_n, Y_n))$ taken from the bivariate normally distributed random variable (X, Y). Therefore, there is a linear relation between the conditional mean of Y and the realization of X, i.e.,

$$\mathsf{E}(Y|\{X = x\}) = \beta_0 + \beta_1 x,$$

for some constants β_0 and β_1.

For each pair of observations (y_i, x_i), we can now write

$$y_i = \mathsf{E}(Y|X = x_i) + e_i = \beta_0 + \beta_1 x_i + e_i, \tag{9.2}$$

where we think of e_i as the realization of a random **error term** ε. In practice, we usually start by posing (9.2) and consider the error term ε to be the random variable. For example, in light of the previous discussion, we could assume that

$$\varepsilon \sim \mathsf{N}(0, \sigma^2). \tag{9.3}$$

Combining (9.2) and (9.3) with the assumption that X and ε are independent gives us what we call the **normal linear regression model**:

$$Y_i = \beta_0 + \beta_1 X_i + \varepsilon_i, \quad (\varepsilon_i)_{i=1}^n \overset{iid}{\sim} \mathsf{N}(0, \sigma^2).$$

9.2   Estimation: Least squares method

The first task that we face is to estimate the parameters β_0 and β_1. As with several models before, we treat σ as a nuisance parameter. We will develop the method that has historically been used most often, the so-called method of **least squares**. The basic idea behind it is as follows. Suppose that the data are plotted on a scatter plot through which we draw a straight line,

$$y = b_0 + b_1 x,$$

for a particular choice of b_0 and b_1. When evaluated at a particular x this line gives a prediction for $\mathsf{E}(Y|X = x)$, namely, $b_0 + b_1 x$. We can define the difference between the actual outcome y_i and this prediction by

$$e_i = y_i - (b_0 - b_1 x_i).$$

The goal is to find the straight line (i.e., values for the intercept b_0 and slope b_1) which "best fits" the data. See Figure 9.2 for a graphical interpretation, where the error of the eighth observation from Example 9.1 is indicated.

We can define "best fit" in many ways, but typically we mean the line that minimizes the sum of squared errors. So, if you define the quantity

$$\mathsf{SSE} := \sum_{i=1}^{n}(y_i - \beta_0 - \beta_1 x_i)^2,$$

then the **ordinary least squares** (OLS) estimators for β_0 and β_1 are given by the values $\hat{\beta}_0^{OLS}$ and $\hat{\beta}_1^{OLS}$ that minimize SSE. From the theory of optimization, it follows that $\hat{\beta}_0^{OLS}$ and $\hat{\beta}_1^{OLS}$ should satisfy the first-order conditions

$$\frac{\partial \mathsf{SSE}}{\partial \beta_0} = 0, \quad \text{and} \quad \frac{\partial \mathsf{SSE}}{\partial \beta_1} = 0.$$

[We should technically also check the second-order conditions, but we will ignore those here.]

Solving this system of two equations in two unknowns, we find that

$$\hat{\beta}_1^{OLS} = \frac{\sum_{i=1}^{n}(X_i - \bar{X})(Yi - \bar{Y})}{\sum_{i=1}^{n}(X_i - \bar{X})^2} = \frac{S_{xy}}{S_x^2}, \quad \text{and} \quad \hat{\beta}_0^{OLS} = \bar{Y} - \bar{X}\hat{\beta}_1^{OLS},$$

where S_{xy} denotes the **sample covariance**,

$$S_{xy} = \frac{1}{n-1} \sum_{i=1}^{n}(X_i - \bar{X})(Y_i - \bar{Y}).$$

Comparing $\hat{\beta}_0^{OLS}$ and $\hat{\beta}_1^{OLS}$ to β_0 and β_1 as given in (9.1), we see that they are the sample analogues of the (population) parameters β_0 and β_1.

[Note that we didn't actually need the distributional assumptions to derive these estimators. For those of you familiar with linear algebra, the OLS estimators determine the orthogonal projection of the vector $(y_1, \ldots, y_n)$ on the span of $(x_1, \ldots, x_n)$. Of course, in order to derive properties of the OLS estimators and to conduct inferences, we *will* use the distributional assumptions.]

Example 9.2 (Example 9.1 continued). For the data in Example 9.1, the estimated coefficients are

$$\hat{\beta}_0^{OLS} = 35.99, \quad \text{and} \quad \hat{\beta}_1^{OLS} = 0.43.$$

So, an additional mark on the midterm exam is, on average, related to an additional 0.43 marks in the end-of-term exam. The estimation indicates that we would predict someone who scores 0 marks in the midterm to obtain 35.99 marks in the end-of-term exam. This is perhaps somewhat unrealistic, but is due to the fact that we are assuming that the conditional mean of the end-of-term exam score is a linear function of the midterm score. Also note that $\hat{\beta}_1^{OLS} > 0$, which means that performance in the end-of-term exam is, on average, better than performance in the midterm. It also implies that students with higher midterm marks, on average, have a higher difference in marks than students with lower midterm results. The line

$$\hat{y} = \hat{\beta}_0^{OLS} + \hat{\beta}_1^{OLS} x, \tag{9.4}$$

is plotted in Figure 9.2. ◁

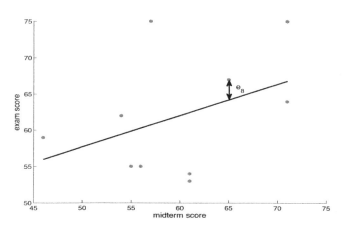

FIGURE 9.2: Scatter plot of midterm and exam scores with fitted OLS regression line.

As an aside, you may wonder if we could estimate β_0 and β_1 using the maximum likelihood principle (or method of moments) that we studied in Chapter 6. You are asked to derive the ML estimators in Exercise 9.1, where you will see that this is indeed possible and that they give the same estimators as the OLS approach. In fact, the method of moments also leads to the OLS estimators (after choosing appropriate moment conditions, of course).

9.3   Decomposition of errors

The predictions following from (9.4) differ from the observations. The errors made by the OLS line, which we call the **residuals**, are

$$e_i = y_i - \hat{y}_i, \quad i = 1, \ldots, n.$$

We can therefore write

$$y_i = \hat{\beta}_0^{OLS} + \hat{\beta}_1^{OLS} x_i + e_i, \quad i = 1, 2, \ldots, n.$$

You can easily show that (i)

$$\sum_{i=1}^n e_i = 0, \quad \text{and thus that} \quad \sum_{i=1}^n y_i = \sum_{i=1}^n \hat{y}_i,$$

and (ii)

$$\sum_{i=1}^n (y_i - \bar{y})^2 = \sum_{i=1}^n (\hat{y}_i - \bar{y})^2 + \sum_{i=1}^n e_i^2.$$

The quantity on the left-hand side of the final equation is called the **total sum of squares** (TSS). The first quantity on the right-hand side is the **explained sum of squares** (ESS), i.e., it is the variation in the predicted (or explained) values. The second quantity is the already familiar sum of squared errors (SSE), i.e., the variation in the error terms. The relation can thus be written as

$$\mathsf{TSS} = \mathsf{ESS} + \mathsf{SSE}.$$

We may want to measure how successfully the fitted line is capturing the variation in the data. Of course, SSE is minimized by construction. However, it could be small simply because there is not much variation in the data. To correct for this, we can divide by TSS to find that

$$1 = \frac{\mathsf{ESS}}{\mathsf{TSS}} + \frac{\mathsf{SSE}}{\mathsf{TSS}}.$$

From this we can define the **coefficient of determination**, or R^2, as

$$R^2 := \frac{\mathsf{ESS}}{\mathsf{TSS}} = 1 - \frac{\mathsf{SSE}}{\mathsf{TSS}}.$$

By definition, $\mathsf{SSE} \geq 0$, so $0 \leq \mathsf{ESS} \leq \mathsf{TSS}$ and hence

$$0 \leq R^2 \leq 1.$$

If $R^2 = 1$, then the line that we have estimated is a perfect fit, i.e., the data in our scatter plot all lie perfectly on the line $\hat{\beta}_0^{OLS} + \hat{\beta}_1^{OLS} x_i$, since in that case $e_i = 0$ for every i. If $R^2 < 1$, then part of the variability in Y is not explained by the variation in X. In general, the higher R^2, the more of the variability in Y is explained by the variability in X.

9.4 Inferences based on the OLS estimator

In order to draw inferences on β_0 and β_1, such as confidence intervals, hypothesis tests, and p-values, based on $\hat{\beta}_0^{OLS}$ and $\hat{\beta}_1^{OLS}$, respectively, we need to know the sampling distributions of these estimators. First of all, since $\bar{Y}$ follows a normal distribution, and $\hat{\beta}_0^{OLS}$ and $\hat{\beta}_1^{OLS}$ are linear functions of $\bar{Y}$, these estimators will also follow a normal distribution. Since $\hat{\beta}_0^{OLS}$ and $\hat{\beta}_1^{OLS}$ are unbiased estimators of β_0 and β_1, respectively (see Exercise 9.2), it holds that

$$\hat{\beta}_0^{OLS} \sim \mathsf{N}(\beta_0, \mathsf{Var}(\hat{\beta}_0^{OLS})), \quad \text{and} \quad \hat{\beta}_1^{OLS} \sim \mathsf{N}(\beta_1, \mathsf{Var}(\hat{\beta}_1^{OLS})).$$

In Exercise 9.2 you are asked to verify that

$$\mathsf{Var}(\hat{\beta}_0^{OLS}) = \frac{\sigma^2}{n} + \bar{X}^2 \mathsf{Var}(\hat{\beta}_1^{OLS}),$$

$$\mathsf{Var}(\hat{\beta}_1^{OLS}) = \frac{\sigma^2}{\sum_{i=1}^n (X_i - \bar{X})^2} = \frac{\sigma^2}{(n-1)S_X^2},$$

and

$$\mathsf{Cov}(\hat{\beta}_0^{OLS}, \hat{\beta}_1^{OLS}) = -\bar{X} \mathsf{Var}(\hat{\beta}_1^{OLS}).$$

[Note that all these distributions and moments are conditional on $(X_1, \ldots, X_n)$. In order to keep the notation tractable, I typically omit making this precise.]

The problem that we now encounter is that the (conditional) distributions of $\hat{\beta}_0^{OLS}$ and $\hat{\beta}_1^{OLS}$ depend on the standard deviation of the error term, σ. This parameter is typically not known. Therefore, we have to replace σ by an estimator. Recall from Chapters 6 and 8 that, when $(X_i)_{i=1}^n \overset{iid}{\sim} \mathsf{N}(\mu, \sigma^2)$ and we replace σ by the unbiased estimator S_X, we can use the t-distribution. So, to make the distributions of $\hat{\beta}_0^{OLS}$ and $\hat{\beta}_1^{OLS}$ usable for inference, we try to follow this approach here and replace σ by an unbiased estimator. We take the estimator[1]

$$S^2 := \frac{1}{n-2}\sum_{i=1}^{n}\varepsilon_i^2 = \frac{\mathsf{SSE}}{n-2}.$$

As a consequence, the *estimated* variances of $\hat{\beta}_0^{OLS}$ and $\hat{\beta}_1^{OLS}$ are

$$\widehat{\mathsf{Var}}(\hat{\beta}_0^{OLS}) = \frac{S^2}{n} + \bar{X}^2\widehat{\mathsf{Var}}(\hat{\beta}_0^{OLS}), \quad \text{and}$$

$$\widehat{\mathsf{Var}}(\hat{\beta}_1^{OLS}) = \frac{S^2}{\sum_{i=1}^{n}(X_i - \bar{X})^2},$$

and the corresponding **standard errors** (this is what we call the estimated standard deviations of the OLS estimators) are

$$\mathsf{SE}(\hat{\beta}_0^{OLS}) = \sqrt{\frac{S^2}{n} + \bar{x}^2\widehat{\mathsf{Var}}(\hat{\beta}_1^{OLS})}, \quad \text{and}$$

$$\mathsf{SE}(\hat{\beta}_1^{OLS}) = \sqrt{\frac{S^2}{\sum_{i=1}^{n}(x_i - \bar{x})^2}}.$$

It now follows that we can introduce the following random variables:

$$T_0 := \frac{\hat{\beta}_0^{OLS} - \beta_0}{\mathsf{SE}(\hat{\beta}_0^{OLS})} \sim t_{n-2} \quad \text{and} \quad T_1 := \frac{\hat{\beta}_1^{OLS} - \beta_1}{\mathsf{SE}(\hat{\beta}_1^{OLS})} \sim t_{n-2}. \qquad (9.5)$$

[See also Exercise 9.3.] Again we see that, by replacing the variance with an unbiased estimator, the distribution changes from normal to Student's t.

Now we are finally ready to make inferences about β_0 and β_1. Let's start by deriving confidence intervals.

For a chosen confidence coefficient $1-\alpha$, we can use the standard procedure to see that $1 - \alpha$ confidence intervals for β_0 and β_1 are

$$\hat{\beta}_0^{OLS} \pm t_{n-2;1-\alpha/2}\mathsf{SE}(\hat{\beta}_0^{OLS}) \quad \text{and} \quad \hat{\beta}_1^{OLS} \pm t_{n-2,1-\alpha/2}\mathsf{SE}(\hat{\beta}_1^{OLS}),$$

respectively. As a check of your understanding of the construction of confidence intervals, derive this confidence interval yourself using the procedure used in Chapter 7.

Suppose that we wish to test the null hypothesis $H_0 : \beta_1 = \bar{\beta}_1$, where $\bar{\beta}_1$ is a pre-specified constant. Such tests can be based on the fact that, when the null hypothesis is true, it holds that

$$T_1 = \frac{\hat{\beta}_1^{OLS} - \bar{\beta}_1}{\mathsf{SE}(\hat{\beta}_1^{OLS})} \sim t_{n-2}.$$

---

[1]Note that we divide by $n - 2$ rather than by $n - 1$ as previously. This is, roughly speaking, because we now have *two* parameters, β_0 and β_1.

Similarly, we can perform tests with the null hypothesis $H_0 : \beta_0 = \bar{\beta}_0$, where $\bar{\beta}_0$ is a pre-specified constant. Such tests can be based on the fact that, when the null hypothesis is true,

$$T_0 = \frac{\hat{\beta}_0^{OLS} - \bar{\beta}_0}{\mathsf{SE}(\hat{\beta}_0^{OLS})} \sim t_{n-2}.$$

One-sided and two-sided tests are then carried out exactly as explained in Chapter 8.

9.5 Linear regression, causation, and correlation

Researchers often want to know whether a dependent variable x "causes" a variable y. In our example of midterm and end-of-term exams, it is very tempting to try and use regression results to answer the question: "do higher midterm results *cause* higher end-of-term results?" You could try to answer such questions by testing $H_0 : \beta_1 = 0$ against $H_1 : \beta_1 \neq 0$. For applied researchers, H_0 is the hypothesis that "x has no influence on y." Rejection of this hypothesis is then interpreted as evidence that "x is significantly correlated with y." Sometimes this is (incorrectly) interpreted as "x is important in determining y."[2]

To see why this test only says something about correlation and not about causation, consider the correlation coefficient

$$\rho = \frac{\sigma_{XY}}{\sigma_X \sigma_Y}.$$

The most obvious estimator for ρ is its sample analogue, denoted by r:

$$r := \frac{S_{XY}}{S_X S_Y}.$$

It can be shown (but that is beyond the scope of this book) that, in order to test $H_0 : \rho = 0$ against $H_1 : \rho \neq 0$, under the null hypothesis it holds that

$$T_r := (n-2)\frac{r^2}{1-r^2} \sim F_{1,n-2},$$

i.e., T_r follows an F-distribution with 1 and $n-2$ degrees of freedom (see Appendix C.3).

---

[2]Performing the aforementioned test is easy, because under this null,

$$T_1 = \frac{\hat{\beta}_1^{OLS}}{\mathsf{SE}(\hat{\beta}_1^{OLS})} \sim t_{n-2}.$$

Now, look again at T_1. Convince yourself that

$$T_1^2 = \left(\frac{\hat{\beta}_1^{OLS}}{\mathsf{SE}(\hat{\beta}_1^{OLS})}\right)^2 = (n-2)\frac{R^2}{1-R^2} = (n-2)\frac{r^2}{1-r^2}.$$

So, testing $H_0 : \beta_1 = 0$ against $H_1 : \beta_1 \neq 0$ is the same as testing whether Y and X are correlated or not. Therefore, the test $H_0 : \beta_1 = 0$ against $H_1 : \beta_1 \neq 0$ is a test of correlation, not causation, and as should be obvious:

| Correlation does not necessarily imply causation. |

9.6 Chapter summary

We developed the normal linear regression model, in which we postulate that, conditional on X, the variable Y has an expectation that is linear in X:

$$\mathsf{E}(Y|X = x) = \beta_0 + \beta_1 x.$$

We built the normal linear regression which stipulates that

1. $Y_i = \beta_0 + \beta_1 X_i + \varepsilon_i$, $i = 1, \ldots, n$,

2. $(\varepsilon_i)_{i=1}^n \overset{iid}{\sim} \mathsf{N}(0, \sigma^2)$,

3. X_i and ε_i are independent.

We developed the least squares method to estimate β_0 and β_1.

By estimating the error standard deviation σ by the unbiased estimator S, we derived the distributions of $\hat{\beta}_0^{OLS}$ and $\hat{\beta}_1^{OLS}$:

$$T_0 = \frac{\hat{\beta}_0^{OLS} - \beta_0}{\mathsf{SE}(\hat{\beta}_0^{OLS})} \sim t_{n-2}, \quad \text{and} \quad T_1 = \frac{\hat{\beta}_1^{OLS} - \beta_1}{\mathsf{SE}(\hat{\beta}_1^{OLS})} \sim t_{n-2},$$

where

$$\mathsf{SE}(\hat{\beta}_0^{OLS}) = \sqrt{\frac{S^2}{n} + \bar{X}^2 \frac{S^2}{\sum_{i=1}^n (X_i - \bar{X})^2}}, \quad \text{and}$$

$$\mathsf{SE}(\hat{\beta}_1^{OLS}) = \sqrt{\frac{S^2}{\sum_{i=1}^n (X_i - \bar{X})^2}}.$$

The statistics T_0 and T_1 can then be used to derive confidence intervals and hypothesis tests in the usual way. In particular, we saw that the test $H_0 : \beta_1 = 0$ against $H_1 : \beta_1 \neq 0$ is equivalent to the test $H_0 : \rho = 0$ against $H_1 : \rho \neq 0$.

9.7 Exercises, problems, and discussion

Exercises

Exercise 9.1. Consider the normal linear regression model

$$Y_i = \beta_0 + \beta_1 X_i + \varepsilon_i, \quad \text{with} \quad (\varepsilon_i)_{i=1}^n \overset{iid}{\sim} N(0, \sigma^2),$$

and X_i and ε_i independent. For the purpose of this exercise, you should treat $(x_1, \ldots, x_n)$ as non-random. That is, the independent variable is not considered to be the realization of a random variable, but rather a given number.

(a) Derive the distribution of Y.

(b) Derive the likelihood function for β_0 and β_1.

(c*) Show that $\hat{\beta}_{0,ML} = \hat{\beta}_0^{OLS}$ and $\hat{\beta}_{1,ML} = \hat{\beta}_1^{OLS}$.

Exercise 9.2 (*). Consider the normal linear regression model.

(a) Show that $\hat{\beta}_0^{OLS}$ and $\hat{\beta}_1^{OLS}$ are unbiased estimators of β_0 and β_1, respectively.

(b) Compute $\text{Var}(\hat{\beta}_1^{OLS})$.

(c) Use (b) to compute $\text{Var}(\hat{\beta}_0^{OLS})$.

Exercise 9.3 (*). Define $S^2 = \frac{1}{n-2} \sum_{i=1}^n e_i^2$.

(a) Show that S^2 is an unbiased estimator for σ^2.

(b) Show that

$$\frac{\hat{\beta}_0^{OLS} - \beta_0}{\sqrt{\text{Var}(\hat{\beta}_0^{OLS})}} \sim N(0,1) \quad \text{and} \quad \frac{\hat{\beta}_1^{OLS} - \beta_1}{\sqrt{\text{Var}(\hat{\beta}_1^{OLS})}} \sim N(0,1).$$

(c) Show that

$$\frac{\hat{\beta}_0^{OLS} - \beta_0}{\text{SE}(\hat{\beta}_0^{OLS})} \sim t_{n-2} \quad \text{and} \quad \frac{\hat{\beta}_1^{OLS} - \beta_1}{\text{SE}(\hat{\beta}_1^{OLS})} \sim t_{n-2}.$$

Exercise 9.4. Consider the normal linear regression model. Use (9.5) and the procedure used in Chapter 7 to derive $1 - \alpha$ confidence intervals for β_0 and β_1.

Exercise 9.5 (regression towards the mean). Consider the normal linear regression model. Suppose that, conditional on $\{X = x\}$, you predict the value for Y to be given by

$$\hat{y} = \hat{\beta}_0^{OLS} + \hat{\beta}_1^{OLS} x.$$

(a) Show that, if x is u units of standard deviation from $\bar{x}$, then $\hat{y}$ is $r \cdot u$ units of standard deviation from $\bar{y}$, where r is the sample correlation coefficient.

(b) Suppose that X denotes a father's height (in m) and Y denotes a son's height (in m). Assume that $r = 0.5$. If a father is 3 units of standard deviation taller than the mean, how much taller than the mean do you predict the son to be?

(c) Historically this phenomenon is called **regression towards the mean**. Why is this an appropriate term?

Problems

Problem 9.1. A researcher has data from 66 households and is interested in the relationship between weekly household consumption Y and weekly household income X. These data are recorded as (y_i, x_i) for $i = 1, 2, \ldots, 66$. The results of an OLS estimation of the normal linear regression model are

$$y_i = \underset{(1.0)}{5.0} + \underset{(0.03)}{0.24}\, x_i + e_i, \quad R^2 = 0.5, \quad n = 66,$$

where e_i denotes a residual and the figures in round brackets are the estimated standard errors.

(a) Comment on the interpretation of the point estimates of β_0 and β_1.

(b) Compute 95% confidence intervals for β_0 and β_1.

(c) It is claimed that households consume approximately one quarter of any additional income. Test this claim at the 5% level.

(d) Test the hypothesis of no correlation between household consumption and income at the 5% level, both using the statistic T_1 and T_r.

(e∗) Give an economic reason why the relation between income and consumption might not be linear.

Problem 9.2. You want to investigate the relation between the size of initial public offerings (IPOs) and the fees that investment banks charge for executing them. IPO size is measured in terms of 10,000 shares placed, whereas the fee is measured as a % value of the total IPO value. The fee can thus be thought of as a commission on sales revenue. For 10 IPOs you have data as described in Table 9.2.

(a) Formulate an appropriate normal linear regression model and find the OLS estimates for the intercept and slope. Also compute R^2.

You are disappointed, because you feel that you need $R^2 \geq 0.8$ to have any chance of publishing your results. When talking about your frustrations to an economist friend, she says the following.

Actually, I don't think there will be a linear relationship between IPO size and fee commanded. Small IPOs are typical for small firms, which have less bargaining power than the big investment banks which can execute the IPO. So, the investment bank can command a high fee. Not too high of course, because otherwise competition will kick in. Say, a maximum fee of 20%. Also, for a small IPO the relative fee must be quite high for the investment bank to recoup its fixed costs, as there is a minimum (legal and administrative) effort required for any IPO. For bigger IPOs the bargaining power of the investment banks is reduced and the average fixed costs of the IPO also go down. This suggests that fees go down after a while to a level of, say, a minimum of 2%. So, I would postulate a relationship between IPO size ($SIZE$) and fee (FEE) that looks something like

$$FEE = \frac{1}{5 + \gamma_1 SIZE^{\gamma_2}}. \tag{9.6}$$

(b) Draw a scatter plot of the data and draw the graph of (9.6) for $\gamma_1 = 6$ and $\gamma_2 = 3$.

(c) Rewrite (9.6) as a linear equation, using a logarithmic transformation. Now formulate a new linear regression model. How do you need to transform the data? How are the intercept and slope linked to γ_1 and γ_2?

(d) Compute the OLS estimates for the intercept and slope and use these to find estimates for γ_1 and γ_2. Also compute the R^2 of this regression.

You are so pleased that you now find $R^2 \geq 0.8$ that you immediately run to your local artisan chocolatier to buy your economist friend some chocolates. On your way to the shop you meet another friend (a statistician), whom you regale with stories of your empirical triumph. Your statistician friend listens with increasing skepticism and finally only manages to mutter:

I'm not very convinced by your using an 0.8 threshold for R^2 in the second regression. But if you think it's appropriate....

(e) Why is your friend so skeptical? Does she think you should have used a higher/lower cut-off for R^2? Why? What experiment does your friend think you've conducted (and should thus have modeled)?

Problem 9.3 ($*$). The capital asset pricing model (CAPM) is a cornerstone of the theory of finance, which predicts that the returns on portfolios of assets in excess of the risk-free rate, r_f, are correlated only with the excess return on the so-called "market portfolio."[3] If R_p denotes the (random) return on some

---

[3]If you are unfamiliar with this terminology, it suffices for our purpose here to think of the S&P500, the FTSE100, or some other broad market index.

TABLE 9.2: Data on 10 IPOs, recording size (in 10,000 shares) and fee (as a % of IPO value).

Size	Fee (%)	Size	Fee (%)
1.6136	2.00	0.5214	15.11
1.7672	2.00	3.0714	2.00
1.0190	5.62	0.8088	12.45
2.0209	3.41	2.2771	2.00
1.5179	4.03	3.0002	2.00

portfolio p, this implies that

$$R_p - R_f = \beta(R_m - r_f),$$

where R_m denotes the return on the market portfolio.

A fund manager's over-performance or under-performance can be measured by deviations from this equation. In order to judge the performance of a fund manager, it is common practice to consider the regression

$$R_{p,i} - r_f = \alpha + \beta(R_{m,i} - r_f) + \varepsilon_i, \tag{9.7}$$

where $(\varepsilon_i)_{i=1}^n \overset{iid}{\sim} \mathsf{N}(0, \sigma^2)$, based on monthly data. So, we use a normal linear regression model with $Y = R_{p,i} - r_f$ and $X = R_{m,i} - r_f$. If the CAPM is correct, then $\alpha = 0$.

(a) Portfolio managers' bonuses are often based on their "outperforming the market" (also called "creating Alpha"). For the regression in (9.7), formulate an appropriate null and alternative hypothesis to test if the manager outperforms the market.

Obviously, the manger's performance may be due to good or bad luck, rather than skill. Therefore, a manager will get a bonus only if the test you designed in (a) provides evidence at, say, the 5% level, that the manager has outperformed the market. For a particular manager, suppose that the result of the regression (9.7) is

$$R_{p,i} - r_f = \underset{(0.3602)}{0.2\%} + \underset{(0.0689)}{1.2} (R_{m,i} - r_f) + e_i,$$

where e_i denotes a residual and the figures in round brackets are the estimated standard errors. In addition, it is given that $s = 2\%$ and $\bar{x} = 1\%$.

(b) Use the expression for $\mathsf{SE}(\hat{a}^{OLS})$ to show that the sample size is $n = 32$.

(c) Show that $R^2 = 0.91$.

(d) Perform the test you designed in (a) at the 5% level. Did this manager statistically significantly outperform the market?

(e) How many years worth of observations do you need if the test conducted in part (d) is required to have a power of 85% against $\alpha = 0.5\%$. [Hint: use the approximations $\mathsf{SE}(\hat{\alpha}^{OLS}) \approx \hat{S}/\sqrt{n}$ and $t_{n-2;1-\alpha} \approx z_{1-\alpha}$ for large n.]

Discussion

Discussion 9.1. The (frequentist) theory of statistical inference as explored so far starts from a mathematical model of a random experiment. Social scientists often do not have experimental, but **observational** data. That is, our data are observations that are not the result of a carefully designed experiment. The way we work is often the opposite of the natural scientist. Their approach is typically as follows.

1. Formulate a theory.

2. Design an experiment to test the theory.

3. State statistical hypotheses.

4. Formulate a statistical model, design and conduct a NPT.

What social scientists often do is something like the following.

1. Formulate a theory.

2. Find data relevant to your problem.

3. State statistical hypotheses.

4. Formulate several statistical models, design NPTs and report the statistically significant results.

(a) Leamer (1978) calls the social scientist's approach "Sherlock Holmes inference." Why do you think he uses that term?

(b) Discuss the dangers of using (frequentist) statistical methods in the way described above.

(c) Many social scientists are now designing laboratory experiments or try to find data that are the result of what are called **natural experiments**. Find some information on these data gathering methods on the internet and discuss why they are valuable to social scientists.

Chapter 10

Bayesian Inference

> When I first learned a little statistics, I felt confused, [...] because I found it difficult to follow the logic by which inferences were arrived at from data. [...] It appeared that the books I looked at were not asking the questions that would naturally occur [...], and that instead they answered some rather recondite questions which no one was likely to ask. [...] I was told that there was another theory of statistical inference, [...and] that this theory could lead to the sorts of conclusions that I had naïvely expected to get from statistics[.]
>
> Lee (2012, pp. xxi, xxii)

10.1 Introduction

One of the reasons many students find statistical inference a difficult subject is the way in which probability plays a role: it is crucial in determining formulae for confidence intervals and hypothesis tests, but *cannot* be used to refer to the outcomes/realizations of such procedures. In the context of the "two worlds" analogy: these probability statements are about the sample world, not the real world. The reason for this is that the parameter θ about which inferences are made is viewed as a constant. We may not know the precise value of this constant, but it is a constant nonetheless. If you look carefully at the arguments in Chapters 7 and 8, you will see that all probability statements are based on the sampling distribution of a statistic, *conditional on the unknown parameter*. The p-value of $H_0 : \theta = \theta_0$ versus $H_1 : \theta > \theta_0$, for example, is computed as $\mathsf{pval} = \mathsf{P}(T > t | \theta = \theta_0)$, for some test statistic T and its observed realization t. So, we view the parameter as constant and the sample as a draw from a random variable.

The frequentist approach is based on the view that nature determines θ and that we, statisticians, are trying to say something about it. Another approach would be to take a more subjective view: as a statistician I am uncertain about the value of the parameter θ and should therefore model it as a random variable. This implies that θ should have a distribution, so that I can meaningfully speak about, say, "the probability that θ falls between a and *b*." In the "two worlds" analogy: such probability statements *are* statements about the real world.

The branch of statistics that takes this view is called **Bayesian statistics** (named after the same chap as Bayes' rule in Chapter 2). It is a very active and important field of statistical research, both theoretically and applied. I do not have space here to develop anything more than the basic ideas through a simple statistical model, but for an excellent textbook-length treatment of the subject, see Lee (2012).

Before we get stuck in the details, it makes sense to try and keep in mind the main difference between the frequentist and Bayesian points of view. Bowley (1937, p. 409) used a nice analogy with target shooting:

> [Frequentists study the problem:] given the target how will shots be dispersed? [In contrast, Bayesians study the problem:] given the shot-marks, what was the target?

10.2   Basic concepts

Suppose that you have a statistical model of a random sample that is drawn from a random variable with density (or mass function) f_θ, where θ is the unknown parameter. Since Bayesians view θ as the realization of a random variable, we need to give it a distribution. In order to distinguish between the parameter as a random variable and its possible realizations, I follow Lee (2012) by denoting the parameter as a random variable using a tilde. Thus, $\tilde{\theta}$ is a random variable and θ denotes one of its possible realizations. All possible realizations are assumed to lie in a set $\Theta \subset \mathbb{R}$. Let's denote the density (or mass function) of the random variable $\tilde{\theta}$ by p. This distribution has to be chosen by the modeler before the analysis starts and is thus part of the statistical model. Choosing an appropriate distribution is part of the (M) bit in the MAD procedure. This distribution is called the **prior distribution**.

The idea is to use the data to get a more accurate description of the parameter $\tilde{\theta}$. This is achieved by computing the distribution of $\tilde{\theta}$, *conditional* on the data $(x_1, \ldots, x_n)$. To save on notation, denote the vector of observations[1] by $\boldsymbol{x}$, i.e., $\boldsymbol{x} := (x_1, \ldots, x_n)$. Denote the density (or mass function) of the distribution of $\tilde{\theta}$ conditional on $\boldsymbol{x}$ by $p(\cdot|\boldsymbol{x})$ and call it the **posterior distribution**. Finally, denote (with some abuse of notation) the density (or mass function) of the random sample by $f(\cdot|\theta)$, i.e.,

$$f(\boldsymbol{x}|\theta) := \prod_{i=1}^{n} f_\theta(x_i).$$

---

[1]Note that we are dealing with the observations here, not the random variables $(X_1, \ldots, X_n)$ of which they are realizations. This is a crucial difference between frequentist and Bayesian methods.

Note that, in the parlance of Section 6.2, this is nothing more than the likelihood function $L(\theta)$. It can be shown that Bayes' rule can be extended to random variables, so that

$$p(\theta|\boldsymbol{x}) = \frac{p(\theta)f(\boldsymbol{x}|\theta)}{\int_\Theta f(\boldsymbol{x}|\theta)d\theta}. \qquad (10.1)$$

The denominator in the above expression no longer depends on the argument, θ, of the posterior, because it has been integrated (or summed) out. So, we can think of the denominator as a constant. This means that we can write that the posterior distribution of $\tilde{\theta}$ is proportional (equal but for a multiplicative constant) to the prior times the likelihood. The mathematical symbol for "proportional to" is $\propto$, so that we can write

$$p(\theta|\boldsymbol{x}) \propto p(\theta)f(\boldsymbol{x}|\theta). \qquad (10.2)$$

That is, the posterior is obtained by multiplying the prior and the likelihood function:

posterior $\propto$ prior $\times$ likelihood.

This formulation tells you to focus on the bit of the posterior distribution that actually depends on the parameter θ. Only afterwards should you worry about the constant that turns $p(\cdot|\boldsymbol{x})$ into a density, i.e., makes it integrate to unity. Until the 1990s a big problem for Bayesian analysis was that the denominator in (10.1) is often very difficult to compute. With the increase of computing power, many methods have been developed to numerically evaluate this integral. Fortunately for applied researchers, these methods have been implemented in user-friendly software so that the practical barriers to using Bayesian analysis are much lower than they once were.[2]

It is the reliance of the Bayesian school of thought on Bayes' rule that explains its name. In a nutshell, the approach uses the following steps:

1. Formulate a distribution for the unknown parameter (the prior).

2. Conditional on the parameter, formulate a distribution for the sample (the likelihood).

3. Use Bayes' rule to combine the prior and the likelihood into a distribution for the unknown parameter that is conditional on the data (the posterior).

In the "two worlds" analogy, the prior is a distribution over the real world, whereas the likelihood is a distribution over the sample world. The use of Bayes' rule turns the prior into another distribution over the real world, after we have used the information obtained from the sample world.

---

[2]Applicability of Bayesian methods has been revolutionized mainly by the advent of simulation methods, in particular, the Markov chain Monte Carlo (MCMC) method. See Lee (2012) or Albert (2009) for details.

10.3   A simple statistical model from a Bayesian perspective

Suppose that we have a statistical model $(X_i)_{i=1}^n \overset{iid}{\sim} \mathsf{N}(\mu, \sigma^2)$, where σ is known. Of course, this is a very restrictive assumption, but it allows us to keep the computations as straightforward as possible. The simple model presented here still conveys the basic ideas, however, and should equip you to appreciate Bayesian analyses if/when you encounter them.

The first task is to complete the model by specifying a prior distribution for the unknown parameter $\tilde{\mu}$. Since our sample is drawn from a normal distribution, there may be an argument to assume a normal distribution for $\tilde{\mu}$ as well: $\tilde{\mu} \sim \mathsf{N}(\mu_0, \sigma_0^2)$, with density

$$p(\mu) \propto \exp\left\{ -\frac{1}{2\sigma_0^2}(\mu - \mu_0)^2 \right\}.$$

The likelihood function (which we already computed in Section 6.2) is

$$f(\boldsymbol{x}|\mu) \propto \exp\left\{ -\frac{1}{2\sigma^2}\sum_{i=1}^n (x_i - \mu)^2 \right\}.$$

Now that we know the prior and likelihood, we can start thinking about the posterior by applying Bayes' rule:

$$p(\mu|\boldsymbol{x}) \propto p(\mu)f(\boldsymbol{x}|\mu)$$

$$\propto \exp\left\{ -\frac{1}{2\sigma_0^2}(\mu - \mu_0)^2 \right\} \exp\left\{ -\frac{1}{2\sigma^2}\sum_{i=1}^n (x_i - \mu)^2 \right\}$$

$$= \exp\left\{ -\frac{1}{2\sigma_0^2}(\mu - \mu_0)^2 - \frac{1}{2\sigma^2}\sum_{i=1}^n (x_i - \mu)^2 \right\}.$$

If we now work out the quadratic terms, interchange the sums, and rearrange a bit, we get

$$p(\mu|\boldsymbol{x}) \propto \exp\left\{ -\frac{1}{2\sigma_0^2}(\mu^2 - 2\mu\mu_0 + \mu_0^2) - \frac{1}{2\sigma^2}\left(\sum_{i=1}^n x_i^2 - 2\mu \sum_{i=1}^n x_i + n\mu^2 \right) \right\}$$

$$= \exp\left\{ -\frac{\mu^2}{2}\left(\frac{1}{\sigma_0^2} + \frac{n}{\sigma^2} \right) + \mu\left(\frac{\mu_0}{\sigma_0^2} + \frac{\sum_{i=1}^n x_i}{\sigma^2} \right) \right\}$$

$$\times \exp\left\{ -\frac{\mu_0^2}{2\sigma_0^2} - \frac{\sum_{i=1}^n x_i^2}{2\sigma^2} \right\}.$$

The last term does not depend on the unknown parameter and can thus be regarded as a multiplicative constant so that we can write

$$p(\mu|\boldsymbol{x}) \propto \exp\left\{-\frac{\mu^2}{2}\left(\frac{1}{\sigma_0^2} + \frac{n}{\sigma^2}\right) + \mu\left(\frac{\mu_0}{\sigma_0^2} + \frac{\sum_{i=1}^{n} x_i}{\sigma^2}\right)\right\}$$

$$= \exp\left\{-\frac{\mu^2}{2}\left(\frac{1}{\sigma_0^2} + \frac{1}{\sigma^2/n}\right) + \mu\left(\frac{\mu_0}{\sigma_0^2} + \frac{\bar{x}}{\sigma^2/n}\right)\right\}.$$

This shows that the posterior of $\tilde{\mu}$ only depends on the data through the sample mean $\bar{x}$. This should not surprise us given that $\bar{x}$ is a sufficient statistic for μ (see Section 5.3).

Now define σ_1 such that

$$\frac{1}{\sigma_1^2} = \frac{1}{\sigma_0^2} + \frac{1}{\sigma^2/n}. \tag{10.3}$$

This allows us to simplify notation:

$$p(\mu|\boldsymbol{x}) \propto \exp\left\{-\frac{\mu^2}{2\sigma_1^2} + \mu\left(\frac{\mu_0}{\sigma_0^2} + \frac{\bar{x}}{\sigma^2/n}\right)\right\}.$$

Now take $1/(2\sigma_1^2)$ out of brackets to get

$$p(\mu|\boldsymbol{x}) \propto \exp\left\{-\frac{1}{2\sigma_1^2}\left[\mu^2 - 2\mu\sigma_1^2\left(\frac{\mu_0}{\sigma_0^2} + \frac{\bar{x}}{\sigma^2/n}\right)\right]\right\}.$$

This is beginning to look like a quadratic form. What is missing is the term

$$\left[\sigma_1^2\left(\frac{\mu_0}{\sigma_0^2} + \frac{\bar{x}}{\sigma^2/n}\right)\right]^2.$$

We can, of course, add this term, but then we also have to subtract it to get

$$p(\mu|\boldsymbol{x}) \propto \exp\left\{-\frac{1}{2\sigma_1^2}\left[\mu^2 - 2\mu\sigma_1^2\left(\frac{\mu_0}{\sigma_0^2} + \frac{\bar{x}}{\sigma^2/n}\right) + \sigma_1^4\left(\frac{\mu_0}{\sigma_0^2} + \frac{\bar{x}}{\sigma^2/n}\right)^2\right]\right\}$$

$$\times \exp\left\{\frac{\sigma_1^2}{2}\left(\frac{\mu_0}{\sigma_0^2} + \frac{\bar{x}}{\sigma^2/n}\right)^2\right\}.$$

Again, the last term does not depend on μ and can be treated as a multiplicative constant:

$$p(\mu|\boldsymbol{x}) \propto \exp\left\{-\frac{1}{2\sigma_1^2}\left[\mu - \sigma_1^2\left(\frac{\mu_0}{\sigma_0^2} + \frac{\bar{x}}{\sigma^2/n}\right)\right]^2\right\}.$$

This looks very much like the density function of a normally distributed random variable with variance σ_1^2 and mean

$$\mu_1 := \sigma_1^2\left(\frac{\mu_0}{\sigma_0^2} + \frac{\bar{x}}{\sigma^2/n}\right). \tag{10.4}$$

Since we know that the density of the posterior distribution of $\tilde{\mu}$ must integrate to unity, this implies that the posterior *is* a normal distribution, i.e.,

$$\tilde{\mu}|\boldsymbol{x} \sim \mathsf{N}(\mu_1, \sigma_1^2). \tag{10.5}$$

A few remarks are in order.

1. The posterior mean is a weighted average of the prior mean and the sample mean, with weights inversely proportional to their respective variances:

$$\mu_1 = \frac{1/\sigma_0^2}{1/\sigma_0^2 + n/\sigma^2}\mu_0 + \frac{n/\sigma^2}{1/\sigma_0^2 + n/\sigma^2}\bar{x}.$$

2. The posterior variance depends on both the variance of the prior and the variance of the sample mean. Also note that σ_1 is decreasing in the sample size. So, the precision of the posterior distribution increases with the sample size (in the sense that there is less spread around the posterior mean).

3. In this case both the prior and the posterior are normal. Whenever the choice of prior distribution leads to a posterior distribution in the same family, statisticians speak of a **conjugate prior**.

4. The approach of looking at the bare essentials of the functional form of the posterior density and then trying to see a well-known distribution in it is very common in the Bayesian approach. It works well here because both the prior and the likelihood are exponentials, which, when multiplied, give another exponential. It is very tempting to let the choice of prior be driven by analytical convenience, but one should resist that temptation in situations where a normal prior is not appropriate.

10.4   Highest density regions

We have now fully determined the distribution of the parameter $\tilde{\mu}$ in light of the observed data $\boldsymbol{x}$. This allows us to compute an analogue to confidence intervals, that is, an interval that contains a fraction $1 - \alpha$ of the probability mass of the posterior distribution of $\tilde{\mu}$. Assuming that $Z \sim \mathsf{N}(0,1)$, we can immediately see, using the familiar steps from Chapter 7, that

$$\begin{aligned} 1 - \alpha &= \mathsf{P}(-z_{1-\alpha/2} < Z < z_{1-\alpha/2}) \\ &= \mathsf{P}\left(\mu_1 - \sigma_1 z_{1-\alpha/2} < \tilde{\mu} < \mu_1 + \sigma_1 z_{1-\alpha/2}\right). \end{aligned} \tag{10.6}$$

Now we can say "in light of the data, the true parameter μ lies in the interval $\mu_1 \pm z_{1-\alpha/2}\sigma_1$ with probability $1 - \alpha$." Notice that, because we are treating the unknown parameter as a random variable, we are now allowed to use the word "probability" in our inference. An interval constructed as in (10.6) is called a $1 - \alpha$ **highest density region** (HDR).

10.5 Hypothesis testing

We can use HDRs also to conduct some two-sided hypothesis tests. Fixing a particular realization for the random variable $\tilde{\mu}$, say $\bar{\mu}$, we could reject the null hypothesis $H_0 : \tilde{\mu} = \bar{\mu}$ against the alternative $H_1 : \tilde{\mu} \neq \bar{\mu}$ if $\bar{\mu}$ does not lie in the $1 - \alpha$ HDR for an appropriately chosen value of α.[3]

Bayesian statisticians typically do not use the language of hypothesis testing, however, because the whole idea of fixing rejection regions to control for Type I errors is steeped in the idea of repeated sampling. Bayesians tend to worry only about the sample that is actually observed. This is obvious in the centrality of the posterior distribution, which is conditional on the realization of the sample. Instead, Bayesians think about the extent to which the data **support** the hypotheses. With that in mind, the above two-sided "test" makes not much sense because we can easily compute that

$$p_0 := \mathsf{P}(\tilde{\mu} = \bar{\mu}|\boldsymbol{x}) = 0, \quad \text{and} \quad p_1 := \mathsf{P}(\tilde{\mu} \neq \bar{\mu}|\boldsymbol{x}) = 1.$$

So, the data do not support the null hypothesis at all.

This should not be a surprise, because we have chosen a model in which μ can take any value. Picking out *one* possible value and then comparing it to *all* other possible values is a rather grotesque way of doing things. An alternative approach would be to think about what constitutes a *significant* (in, for example, the economic sense) deviation from $\bar{\mu}$. Suppose that you decide that any deviation from $\bar{\mu}$ by more than some $\varepsilon > 0$ is to be considered significant. Then you can compute the support p_0 for the hypothesis $H_0 : \tilde{\mu} \in (\bar{\mu} - \varepsilon, \bar{\mu} + \varepsilon)$,

$$p_0 := \mathsf{P}\left(\bar{\mu} - \varepsilon < \tilde{\mu} < \bar{\mu} + \varepsilon\right),$$

as well as the support p_1 for the hypothesis $H_1 : \tilde{\mu} \notin (\bar{\mu} - \varepsilon, \bar{\mu} + \varepsilon)$,

$$p_0 := \mathsf{P}\left(\tilde{\mu} < \bar{\mu} - \varepsilon, \tilde{\mu} > \bar{\mu} + \varepsilon\right),$$

and decide to reject H_0 if $p_1 > p_0$.

Alternatively, you could start your analysis by stating two simple hypotheses $H_0 : \tilde{\mu} = \mu_0$ and $H_1 : \tilde{\mu} = \mu_1$ and consider the parameter space to be restricted to $\{\mu_0, \mu_1\}$. Assume that $\mu_1 > \mu_0$. Of course, you can then no longer use the prior that we used before. Instead we can use the prior

$$p(\mu_0) = \mathsf{P}(\tilde{\mu} = \mu_0) = \pi_0, \quad \text{and} \quad p(\mu_1) = \mathsf{P}(\tilde{\mu} = \mu_1) = \pi_1 = 1 - \pi_0,$$

---

[3]Note that I reflect the Bayesian viewpoint of treating parameters as random variables by writing the hypotheses as statements about realizations of random variables.

for some $\pi_0 \in (0,1)$. This implies that the posterior needs to be recomputed. It is easy to see that the support for H_0 and H_1 are given by

$$p_0 = \mathsf{P}(\tilde{\mu} = \mu_0 | \boldsymbol{x}) \propto \pi_0 \exp \left\{ -\frac{1}{2\sigma^2} \sum_{i=1}^{n} (x_i - \mu_0)^2 \right\}, \quad \text{and}$$

$$p_1 = \mathsf{P}(\tilde{\mu} = \mu_1 | \boldsymbol{x}) \propto \pi_1 \exp \left\{ -\frac{1}{2\sigma^2} \sum_{i=1}^{n} (x_i - \mu_1)^2 \right\},$$

respectively.

The **relative support** of H_1 against H_0, on the data, then equals

$$\begin{aligned}
\frac{p_1}{p_0} &= \frac{\pi_1 \exp\left\{ -\frac{1}{2\sigma^2} \sum_{i=1}^{n} (x_i - \mu_1)^2 \right\}}{\pi_0 \exp\left\{ -\frac{1}{2\sigma^2} \sum_{i=1}^{n} (x_i - \mu_0)^2 \right\}} \\
&= \frac{\pi_1}{\pi_0} \exp \left\{ -\frac{(\mu_1 - \mu_0)}{\sigma^2/n} \left(\frac{1}{2}(\mu_1 + \mu_0) - \bar{x} \right) \right\}.
\end{aligned} \tag{10.7}$$

In moving from the first to the second line in (10.7) I have used the fact that $\mu_1^2 - \mu_0^2 = (\mu_1 - \mu_0)(\mu_1 + \mu_0)$. Note that the first term on the right-hand side is the ratio of the prior probabilities, also known as the **prior odds** of the two hypotheses. By analogy, we call p_1/p_0 the **posterior odds**. The second term on the right-hand side is the likelihood ratio that we encountered in Chapter 8. So, we can see that

> posterior odds = prior odds × likelihood ratio.

Another way of reporting the relative support of H_1 against H_0 is by computing the so-called **Bayes factor**,

$$B := \frac{p_0/p_1}{\pi_0/\pi_1} = \frac{p_0}{p_1} \frac{\pi_1}{\pi_0}.$$

Question 10.1. Why does the first line in (10.7) have an equality sign rather than a $\propto$ sign? What happened to the multiplicative constant of the posterior?

In terms of the sufficient statistic we can rewrite (10.7) as saying that we reject H_0 if

$$\begin{aligned}
&\frac{\pi_1}{\pi_0} \exp \left\{ -\frac{(\mu_1 - \mu_0)}{\sigma^2/n} \left(\frac{1}{2}(\mu_1 + \mu_0) - \bar{x} \right) \right\} > 1 \\
\Longleftrightarrow \ &\exp \left\{ -\frac{(\mu_1 - \mu_0)}{\sigma^2/n} \left(\frac{1}{2}(\mu_1 + \mu_0) - \bar{x} \right) \right\} > \frac{\pi_0}{\pi_1} \\
\Longleftrightarrow \ &-\frac{(\mu_1 - \mu_0)}{\sigma^2/n} \left(\frac{1}{2}(\mu_1 + \mu_0) - \bar{x} \right) > \log \left(\frac{\pi_0}{\pi_1} \right) \\
\Longleftrightarrow \ &\bar{x} > \frac{1}{2}(\mu_1 + \mu_0) - \frac{\sigma^2/\sqrt{n}}{\mu_1 - \mu_0} \log \left(\frac{\pi_0}{\pi_1} \right).
\end{aligned} \tag{10.8}$$

Note that the last step uses the assumption that $\mu_1 > \mu_0$. If $\mu_1 < \mu_0$, you reject H_0 if

$$\bar{x} < \frac{1}{2}(\mu_1 + \mu_0) - \frac{\sigma^2/\sqrt{n}}{\mu_1 - \mu_0} \log\left(\frac{\pi_0}{\pi_1}\right).$$

At this stage I would like to make two remarks.

1. The Bayesian approach makes it very clear that support for rival statistical hypotheses provided by the data is always *relative*. In fact, there is a school of thought that develops inference entirely based on the likelihood ratio alone. See Section 10.6 below.

2. The Bayesian framework is very versatile in terms of the kind of statistical inferences you can draw. This is partly because of the freedom you have in choosing a particular prior and partly because you end up with the *entire distribution* of $\bar{\mu}$ conditional on the data. This gives you a lot of scope to exploit that distribution for a large variety of inferences.

Many frequentist statisticians don't like the Bayesian approach precisely because of the second point. Since you can choose any prior you like, they would accuse a Bayesian statistician of being able to draw any conclusion they like by choosing an "appropriate" prior. Of course, a Bayesian statistician can retort that the frequentist is also using ad hoc procedures: why use a 5% level and not 4.5%? Or 1.2%? Nevertheless, Bayesians have taken some of the criticism seriously and developed ways to construct priors that are, in some sense, "vague." Such priors are called **noninformative priors**. In our normal model you could think that, if you have no prior information about the parameter μ, you could choose a uniform prior:

$$p(\mu) \propto 1.$$

In that case, you consider all possible values for μ possible and equally likely.

But now you have a problem: you can't get a uniform distribution over the entire real line, because it can never integrate to unity. Bayesians call this, therefore, an **improper prior**. It turns out that we don't need to worry too much about this, because the posterior that we obtain *is* a well-defined distribution. To see this we perform similar computations as when we assumed the prior to be normal. Below I use pretty much the same steps as in Section 10.3 (so I'll go through it more quickly):

$$p(\mu|\boldsymbol{x}) \propto 1 \times f(\boldsymbol{x}|\mu)$$

$$= \exp\left\{-\frac{1}{2\sigma^2}\sum_{i=1}^{n}(x_i - \mu)^2\right\}$$

$$\propto \exp\left\{-\frac{1}{2\sigma^2/n}(\mu^2 - 2\mu\bar{x})\right\}$$

$$\propto \exp\left\{-\frac{1}{2\sigma^2/n}(\mu - \bar{x})^2\right\}.$$

This looks like the density of a normally distributed random variable with mean $\bar{x}$ and variance σ^2/n, so that we get

$$\tilde{\mu}|\boldsymbol{x} \sim \mathsf{N}(\bar{x}, \sigma^2/n).$$

Note that this posterior distribution only depends on the observed data, through the sample mean $\bar{x}$, and not on the prior.

We can now compute the support of, for example, the null hypothesis $H_0 : \mu \leq \bar{\mu}$ relative to the hypothesis $H_1 : \mu > \bar{\mu}$, for some $\bar{\mu}$, by computing p_0:

$$p_0 = \mathsf{P}(\tilde{\mu} \leq \bar{\mu}|\bar{x}) = \Phi\left(\frac{\bar{\mu} - \bar{x}}{\sigma/\sqrt{n}}\right).$$

A frequentist statistician computing the p-value for the same hypotheses obtains

$$\mathsf{pval} = \mathsf{P}(\bar{X} > \bar{x}|\mu = \bar{\mu}) = 1 - \Phi\left(\frac{\bar{x} - \bar{\mu}}{\sigma/\sqrt{n}}\right) = \Phi\left(\frac{\bar{\mu} - \bar{x}}{\sigma/\sqrt{n}}\right) = p_0.$$

From the analysis above, with an improper prior, the Bayesian finds the same support for H_0 as the frequentist finds evidence against it. You can see from the computations why this happens: the Bayesian and frequentist condition exactly in opposite ways. The frequentist conditions on the parameter, whereas the Bayesian conditions on the data.

10.6   Evidence and the likelihood principle

Recall the critique of using the p-value for statistical inference by Jeffreys (1961, Section 7.2):

> What the use of [the p-value] implies, therefore, is that a hypothesis that may be true may be rejected because it has not predicted observable results that have not occurred.

The problem that Jeffreys points at is that inferences drawn on the basis of p-values (or NPTs for that matter) depend not only on the observations, but also on possible observations that have not occurred. That is, frequentist inference does not satisfy the **likelihood principle** (Berger, 1985, p. 28):

> In making inferences or decisions about θ after $\boldsymbol{x}$ is observed, all relevant experimental information is contained in the likelihood function for the observed $\boldsymbol{x}$. Furthermore, two likelihood functions contain the same information about θ if they are proportional to each other.

The Bayesian approach satisfies this principle, because any inference is based on the posterior distribution and the posterior depends on the data solely through the likelihood function. Some statisticians, such as, for example, Edwards (1992) and Royall (1997) would go even further and say that inferences can *only* be based on the likelihood function. This leads to the **law of likelihood** (Hacking, 1965, p. 62):[4]

> [The data] supports [simple hypothesis $H_1 : \theta = \theta_1$] better than [simple hypothesis $H_2 : \theta = \theta_2$] if the likelihood of [H_1] exceeds that of [H_2].

Note a few implications of this law.

1. The hypotheses H_1 and H_2 are treated symmetrically; you can easily interchange them (unlike the null and alternative hypotheses in the frequentist approach).

2. Statistical evidence is always *relative*: the data can only support a hypothesis in relation to another hypothesis.

3. The law of likelihood can be formulated in terms of the likelihood ratio Λ: the data support H_1 better than H_2 if, and only if, $\Lambda = L(\theta_1)/L(\theta_2) > 1$.

4. The hypothesis that is at least as well supported by the data as any other hypothesis is $H : \theta = \hat{\theta}_{ML}$.

5. The Bayesian approach satisfies the *principle*, but not the *law* of likelihood. This is because the posterior odds are influenced by the prior odds as well as the likelihood ratio.

In the likelihood approach, the likelihood function, and the likelihood function only, measures the *strength of evidence* in favour of one hypothesis compared to another hypothesis. Royall (1997) argues, for example, that if a statistician is requested to present evidence and no more than that, she should simply plot the likelihood function. In order to make statements about strength of evidence, Royall (1997, Section 1.6.1) suggests using the following canonical example.

Suppose that you have an urn with either 50% white balls and 50% black balls, or white balls only. How many white balls do you need to draw (with replacement) in succession before you would say that you have strong or very strong evidence in favour of the all white hypothesis? The probabilistic model for this experiment is $X \sim \mathsf{Bern}(p)$, where X denotes the colour of a ball that is drawn randomly ($X = 1$ if white, $X = 0$ if black), where p is the probability of a white ball being drawn. The two hypotheses are

$$H_1 : p = 1 \quad \text{and} \quad H_2 : p = 1/2.$$

If you draw n balls and they are all white, then the likelihood ratio is $\Lambda = 2^n$. If you think that three white balls in succession presents strong evidence for

---

[4]It has to be noted that Hacking, in later writings, distanced himself somewhat from the likelihood approach and moved towards a more Bayesian viewpoint.

TABLE 10.1: Number of consecutive white balls drawn in the canonical experiment and corresponding likelihood ratios.

n	3	4	5	6	10
Λ	8	16	32	64	1024

the hypothesis of an all white urn, then your cut-off point for strong evidence is a likelihood ratio of 8. If you think five white balls in succession is very strong evidence for the hypothesis of an all white urn, then your cut-off point for very strong evidence is a likelihood ratio of 32. These cut-off points can be used for every application, because strength of evidence is measured in a relative way. Table 10.1 converts the number of white balls drawn in succession (n) to the corresponding likelihood ratio.

Suppose that you have a sample $(X_i)_{i=1}^n \overset{iid}{\sim} \mathsf{N}(\mu, \sigma^2)$, with σ known. Consider the hypotheses (Royall, 1997, Section 4.4)

$$H_1 : \mu = \mu_1 \quad \text{and} \quad H_2 : \mu = \mu_1 + \delta.$$

Since the hypotheses are treated symmetrically, we can, without loss of generality, assume that $\delta > 0$. We already computed that

$$\Lambda = \frac{\exp\left\{-\frac{1}{2\sigma^2}\sum_{i=1}^n (X_i - \mu_1)^2\right\}}{\exp\left\{-\frac{1}{2\sigma^2}\sum_{i=1}^n (X_i - \mu_2)^2\right\}}. \tag{10.9}$$

Once you have set your likelihood ratio cut-off point k and you observe the data, you can immediately see if you have observed strong evidence in favour of H_1 over H_2 ($\Lambda > k$) or if you have observed strong evidence in favour of H_2 over H_1 ($\Lambda < 1/k$). If you find $1/k < \Lambda < k$, you would conclude that you have found **weak evidence**. Of course, it is possible that you find **misleading evidence**, i.e., $\Lambda > k$ if $\mu = \mu_1 + \delta$ or $\Lambda < 1/k$ if $\mu = \mu_1$. Before you see the data you can compute the probabilities of these events. Note that we can write

$$\Lambda = \exp\left\{-\frac{n\delta}{\sigma^2}\left[\bar{X} - (\mu_1 + \delta/2)\right]\right\}. \tag{10.10}$$

The probability of finding weak evidence if H_1 is correct is

$$\begin{aligned}
W_1 :&= \mathsf{P}\left(\frac{1}{k} < \Lambda < k \,\middle|\, H_1\right) \\
&= \mathsf{P}\left(\mu_1 + \frac{\delta}{2} - \frac{\sigma^2}{n\delta}\log(k) < \bar{X} < \mu_1 + \frac{\delta}{2} + \frac{\sigma^2}{n\delta}\log(k) \,\middle|\, H_1\right) \quad (10.11) \\
&= \Phi\left(\frac{\delta\sqrt{n}}{2\sigma} + \frac{\sigma}{\delta\sqrt{n}}\log(k)\right) - \Phi\left(\frac{\delta\sqrt{n}}{2\sigma} - \frac{\sigma}{\delta\sqrt{n}}\log(k)\right).
\end{aligned}$$

It is easy to show that the probability of finding weak evidence if H_2 is correct, W_2, is the same, i.e., $W_1 = W_2$. The probability of finding misleading evidence if H_2 is true is

$$M_2 := \mathsf{P}(\Lambda > k | H_2)$$

$$= \mathsf{P}\left(\bar{X} > \mu_1 + \frac{\delta}{2} + \frac{\sigma^2}{n\delta}\log(k)\Big| H_2\right) \tag{10.12}$$

$$= 1 - \Phi\left(\frac{\sigma}{\delta\sqrt{n}}\log(k) - \frac{\delta\sqrt{n}}{2\sigma}\right).$$

A similar exercise shows that the probability of finding misleading evidence if H_1 is true is

$$M_1 := \mathsf{P}(\Lambda < 1/k | H_1)$$

$$= \mathsf{P}\left(\bar{X} < \mu_1 + \frac{\delta}{2} - \frac{\sigma^2}{n\delta}\log(k)\Big| H_1\right) \tag{10.13}$$

$$= \Phi\left(\frac{\delta\sqrt{n}}{2\sigma} - \frac{\sigma}{\delta\sqrt{n}}\log(k)\right).$$

10.7 Decision theory

At the end of a typical statistics course many students feel a sense of frustration. They spend all this time learning about ways to interpret evidence in the face of uncertainty, but they don't feel it actually helps them *making decisions*. I don't have space here to do anything but scratch the surface, but if you are interested in what is discussed in this section you can consult Pratt et al. (1995) for an applied, or Berger (1985) for a more theoretical exposition of statistical decision theory. Here I follow the notation used by Berger (1985).

In order to think about the basic ideas, let's consider the normal model $(X_i)_{i=1}^n \overset{iid}{\sim} \mathsf{N}(\mu, \sigma^2)$, with σ known and the hypotheses $H_0 : \mu = \mu_0$ and $H_1 : \mu = \mu_1$, with $\mu_1 > \mu_0$. Recall that a frequentist will reject H_0 at level α if

$$\bar{x} > \mu_0 + z_{1-\alpha}\frac{\sigma}{\sqrt{n}}.$$

A Bayesian will view μ as the realization of a random variable $\tilde{\mu}$ that can take the two values μ_0 and μ_1, and assign a prior $\mathsf{P}(\tilde{\mu} = \mu_0) = \pi_0$ and $\mathsf{P}(\tilde{\mu} = \mu_1) = 1 - \pi_0$. She then computes the relative support of H_1 against H_0 using the posterior odds:

$$\frac{p_1}{p_0} = \frac{\pi_1}{\pi_0}\exp\left\{-\frac{(\mu_1 - \mu_0)}{\sigma^2/n}\left(\frac{1}{2}(\mu_1 + \mu_0) - \bar{x}\right)\right\}.$$

This is where the inferences end; the analysis does not tell you *what to do.* Devising procedures that allow you to use statistical evidence to inform decision making is the realm of (statistical) **decision theory**. What we need is some way to link parameter values, observations from some random experiment, and actions that we can take to *consequences* of these actions. To formalize this, recall that the **parameter space** is denoted by Θ. We observe a sample x drawn from some joint distribution F_θ taking values in some **sample space**, which we'll denote by $\mathscr{X}$. The actions that we can take are listed in a set that we call the **action space** and denote by $\mathscr{A}$.

Example 10.1 (estimation). Suppose that you have a sample $(X_i)_{i=1}^n \overset{iid}{\sim}$ $\mathsf{N}(\theta, \sigma^2)$, with σ known and that you wish to estimate θ. The parameter space is then the set of all possible values that θ can take, which, without any additional information, is $\Theta = \mathbb{R}$. The decision that you take consists of choosing a particular value in Θ that you will call your (point) estimate of μ. So, $\mathscr{A} = \Theta = \mathbb{R}$. Since you have a sample of size n, drawn from a normal distribution, the sample space consists of all n-dimensional vectors, i.e., $\mathscr{X} = \mathbb{R}^n$. $\triangleleft$

The *decision* that you take will, in general, depend on the sample that you observe. What you are looking for, therefore, is a **decision rule** that tells you which action to choose given an observed sample, i.e., a function $\delta : \mathscr{X} \to \mathscr{A}$. The set of all decision rules is denoted by $\mathscr{D}$. The *consequences* of your decision depend on the action that you take and the true value of the parameter. Traditionally, statisticians measure the consequences using a **loss function** that maps combinations of parameter values and actions into (monetary) losses.[5] That is, the loss function is a function $L : \mathscr{A} \times \Theta \to \mathbb{R}$.

Example 10.2 (Example 10.1 continued). Obviously you want to choose your estimate $a \in \mathscr{A}$ as close as possible to the true parameter value θ. So you could take the loss of your estimate to be the distance from your estimate to the true parameter value. This suggests taking

$$L(a, \theta) = (\theta - a)^2.$$

[Note the similarity between this formulation and the idea behind the mean squared error.] $\triangleleft$

Suppose that you have no data available at all beyond your prior p over the parameter (viewed as a random variable) $\tilde{\theta}$. What action would you choose? An obvious choice would be the action that *minimizes* the loss you expect to make, i.e., choose $a \in \mathscr{A}$ to minimize the (prior) **expected loss**:

$$\rho(a) := \mathsf{E}_{p(\theta)}\left[L(a, \tilde{\theta})\right] = \int_\Theta L(a, \theta)p(\theta)d\theta.$$

---

[5]The reason for this is that statisticians are usually most interested in making correct decisions and think of erroneous decisions as leading to losses. An economist would be more interested in, for example, profits resulting from her decisions. For an example, see Section 10.8.

Note that I add a subscript to the expectation operator to remind myself (i) what the random variable is and (ii) which distribution I need to use (prior or posterior).

Example 10.3 (Example 10.1 continued). Suppose that you have the prior distribution

$$\tilde{\theta} \sim \mathsf{N}(\theta_0, \sigma_0^2).$$

We can now compute the expected loss:

$$
\begin{aligned}
\rho(a) = \mathsf{E}_{p(\theta)} \left[L(a, \tilde{\theta}) \right] &= \mathsf{E}_{p(\theta)} \left[(\tilde{\theta} - a)^2 \right] \\
&= \mathsf{E}_{p(\theta)} \left[\tilde{\theta}^2 \right] - 2a \mathsf{E}_{p(\theta)} \left[\tilde{\theta} \right] + a^2 \\
&= \mathsf{Var}_{p(\theta)}(\tilde{\theta}) + \mathsf{E}_{p(\theta)} \left[\tilde{\theta} \right]^2 - 2a \mathsf{E}_{p(\theta)} \left[\tilde{\theta} \right] + a^2 \\
&= \sigma_0^2 + \theta_0^2 - 2a\theta_0 + a^2.
\end{aligned}
$$

Note that a is a constant in the computations above: we fix the action $a \in \mathscr{A}$ and then compute the expected loss. Minimizing this function with respect to a gives

$$a^* := \arg\min_{a \in \mathscr{A}} \rho(a) = \theta_0.$$

So, you should estimate θ by the mean of the prior distribution. ◁

When you observe your sample $\boldsymbol{x}$, you can compute the posterior distribution and you can use this distribution to compute the posterior expected loss, which is usually called the **Bayes expected loss** of action $a \in \mathscr{A}$:

$$\rho(a, \boldsymbol{x}) := \mathsf{E}_{p(\theta|\boldsymbol{x})} \left[L(a, \tilde{\theta}) \right] = \int_{\Theta} L(a, \theta) p(\theta|\boldsymbol{x}) d\theta.$$

For each sample $\boldsymbol{x} \in \mathscr{X}$ this procedure prescribes an action $a \in \mathscr{A}$, called the **Bayes decision**. In formal notation, the Bayes decision is a decision rule δ^* such that, for each $\boldsymbol{x} \in \mathscr{X}$,

$$\delta^*(\boldsymbol{x}) = \arg\min_{a \in \mathscr{A}} \rho(a, \boldsymbol{x}).$$

Example 10.4 (Example 10.1 continued). Recall from Section 10.3 that

$$\tilde{\theta}|\boldsymbol{x} \sim \mathsf{N}(\theta_1(\boldsymbol{x}), \sigma_1^2),$$

where

$$\theta_1(\boldsymbol{x}) = \sigma_1^2 \left(\frac{\mu_0}{\sigma_0^2} + \frac{\bar{x}}{\sigma^2/n} \right), \quad \text{and}$$

$$\frac{1}{\sigma_1^2} = \frac{1}{\sigma_0^2} + \frac{1}{\sigma^2/n}.$$

Using exactly the same procedure as in computing the (prior) expected loss, it should be obvious that the Bayes rule is

$$\delta^*(\boldsymbol{x}) = \theta_1(\boldsymbol{x}).$$

So, based on your sample, you should estimate θ by the mean of the posterior distribution. $\triangleleft$

Example 10.5 (hypothesis testing). Suppose that you have a sample $(X_i)_{i=1}^n \overset{iid}{\sim} \mathsf{N}(\theta, \sigma^2)$, with σ known and that you wish to test

$$H_0 : \theta = \theta_0 \quad \text{against} \quad H_1 : \theta = \theta_1.$$

Suppose your prior is such that

$$p(\theta_1) = p \quad \text{and} \quad p(\theta_0) = 1 - p.$$

Recall that the posterior odds are

$$\frac{p(\theta_1 | \boldsymbol{x})}{p(\theta_0 | \boldsymbol{x})} = \frac{p}{1-p} \exp\left\{ -\frac{(\theta_1 - \theta_0)}{\sigma^2/n} \left(\frac{1}{2}(\theta_1 + \theta_0) - \bar{x} \right) \right\}.$$

Our parameter space in this problem is no longer $\mathbb{R}$, because we are focussing on two values of θ only. Rather, $\Theta = \{\theta_0, \theta_1\}$. The actions that we can take are to accept the null (which we will denote by a_0) or to accept the alternative (which we will denote by a_0), so that the action space is $\mathscr{A} = \{a_0, a_1\}$. [Note that, even though I spent a lot of time in Chapter 8 arguing that you shouldn't accept the null ever, here we *can* speak of accepting the null. That is because we are modelling a decision making process rather than a pure inferential process.] The loss function could reflect our concerns about making Type I and Type II errors. Suppose that their losses are taken to be c and d, respectively. We then get

$$L(a, \theta) = c\mathbb{1}_{\theta=\theta_0, a=a_1} + d\mathbb{1}_{\theta=\theta_1, a=a_0}.$$

The Bayes expected loss of action $a \in \mathscr{A}$ is

$$\begin{aligned}
\rho(a, \boldsymbol{x}) &= \mathsf{E}_{p(\theta|\boldsymbol{x})} \left[L(a, \tilde{\theta}) \right] \\
&= \mathsf{E}_{p(\theta|\boldsymbol{x})} \left[c\mathbb{1}_{\theta=\theta_0, a=a_1} + d\mathbb{1}_{\theta=\theta_1, a=a_0} \right] \\
&= c\mathsf{P}(\tilde{\theta} = \theta_0)\mathbb{1}_{a=a_1} + d\mathsf{P}(\tilde{\theta} = \theta_1)\mathbb{1}_{a=a_0} \\
&= cp(\theta_0, \boldsymbol{x})\mathbb{1}_{a=a_1} + dp(\theta_1, \boldsymbol{x})\mathbb{1}_{a=a_0}.
\end{aligned}$$

You choose a_1 if, and only if,

$$\begin{aligned}
&\rho(a_1, \boldsymbol{x}) < \rho(a_0, \boldsymbol{x}) \\
\Longleftrightarrow \ &cp(\theta_0, \boldsymbol{x}) < dp(\theta_1, \boldsymbol{x}) \\
\Longleftrightarrow \ &\frac{d}{c} \frac{p(\theta_1, \boldsymbol{x})}{p(\theta_0, \boldsymbol{x})} > 1 \\
\Longleftrightarrow \ &\frac{d}{c} \frac{p}{1-p} \exp\left\{ -\frac{(\theta_1 - \theta_0)}{\sigma^2/n} \left(\frac{1}{2}(\theta_1 + \theta_0) - \bar{x} \right) \right\} > 1 \\
\Longleftrightarrow \ &\bar{x} > \frac{1}{2}(\theta_1 + \theta_0) - \frac{\sigma^2/\sqrt{n}}{\theta_1 - \theta_0} \left[\log\left(\frac{1-p}{p} \right) + \log\left(\frac{c}{d} \right) \right],
\end{aligned}$$

where I have followed the same steps as in the derivation of (10.8). [Note that, if $c = d$, we obtain exactly the same rejection rule as in 10.8.] So, the Bayes decision rule is

$$
\delta(\boldsymbol{x}) = \begin{cases} a_1 & \text{if } \bar{x} > \frac{1}{2}(\theta_1 + \theta_0) - \frac{\sigma^2/\sqrt{n}}{\theta_1 - \theta_0}\left[\log\left(\frac{1-p}{p}\right) + \log\left(\frac{c}{d}\right)\right] \\ a_0 & \text{if } \bar{x} < \frac{1}{2}(\theta_1 + \theta_0) - \frac{\sigma^2/\sqrt{n}}{\theta_1 - \theta_0}\left[\log\left(\frac{1-p}{p}\right) + \log\left(\frac{c}{d}\right)\right]. \end{cases}
$$

If

$$
\bar{x} = \frac{1}{2}(\mu_1 + \mu_0) - \frac{\sigma^2/\sqrt{n}}{\mu_1 - \mu_0}\left[\log\left(\frac{1-p}{p}\right) + \log\left(\frac{c}{d}\right)\right],
$$

you are indifferent between both actions and you can choose either or flip a coin.   ◁

10.8   An economic decision problem with statistical information

From an economic point of view, there is something strange about the decision theoretic formulation of statistical inference. Why would we only be interested in penalties for making mistakes? Surely, we should also take into account the positive payoffs we might get if we make the correct decision. In principle, there is nothing stopping you from formulating a loss function that incorporates negative losses for positive gains. For most economic and business applications, it is intuitively more appealing to maximize profits, rather than to minimize losses. This section provides an example.

Suppose that you are the manager of an energy company and you need to decide whether or not to exploit a particular gas field. Suppose that the amount of gas that we can extract from the field is a draw from some random variable with unknown mean. Based on geological information, we may know that the unknown mean can take the values $\theta_0 > 0$ or $\theta_1 > \theta_0$ only. So your parameter space is $\Theta = \{\theta_0, \theta_1\}$. Your prior assessment of these hypotheses is such that $p(\theta_1) = p$. We consider two possible actions: exploit the gas field (denoted by a_1) or don't (a_0), so that $\mathscr{A} = \{a_0, a_1\}$. [So far, the set-up is very similar to Example 10.5.]

Suppose that we can sell the gas at a constant price of \$1 and that the costs of exploiting the field are $I > 0$. This implies that our expected profit equals $\theta_1 - I$ if $\theta = \theta_1$ and $\theta_0 - I$ if $\theta = \theta_0$. To make the problem economically interesting, we assume that $\theta_1 > I > \theta_0$, so that we make a profit (loss) if $\bar{\theta} = \theta_1$ ($\bar{\theta} = \theta_0$). If we don't exploit the field, we don't get any revenues, but we also don't make any costs. We can summarize this payoff or **utility function** $U : \mathscr{A} \times \Theta \to \mathbb{R}$ as in Table 10.2.[6]

---

[6]Note that, in this case, the loss of a Type II error is 0.

TABLE 10.2: Payoff table for a statistical decision problem.

		State of nature	
		$\{\tilde{\theta} = \theta_0\}$	$\{\tilde{\theta} = \theta_1\}$
Decision	don't exploit	0	0
	exploit	$\theta_0 - I$	$\theta_1 - I$

If we now get a sample x of geological measurements from $(X_i)_{i=1}^{n} \overset{iid}{\sim}$ $\mathsf{N}(\theta, \sigma^2)$, we can use the posterior distribution to compute the expected profit, call it E, from exploiting the field:

$$E = p(\theta_1|\boldsymbol{x})(\theta_1 - I) + p(\theta_0|\boldsymbol{x})(\theta_0 - I).$$

In order to compute E we really need to know the posterior probabilities. We do this by combining the fact that

$$\frac{p(\theta_1|\boldsymbol{x})}{p(\theta_0|\boldsymbol{x})} = \frac{p}{1-p} \exp\left\{ -\frac{(\theta_1 - \theta_0)}{\sigma^2/n} \left(\frac{1}{2}(\theta_1 + \theta_0) - \bar{x} \right) \right\},$$

with the observation that $p(\theta_1|\boldsymbol{x}) + p(\theta_0|\boldsymbol{x}) = 1$. Solving the equation

$$\frac{1 - p(\theta_0|\boldsymbol{x})}{p(\theta_0|\boldsymbol{x})} = \frac{p}{1-p} \exp\left\{ -\frac{(\theta_1 - \theta_0)}{\sigma^2/n} \left(\frac{1}{2}(\theta_1 + \theta_0) - \bar{x} \right) \right\}$$

gives that

$$p(\theta_0|\boldsymbol{x}) = \left(\frac{p}{1-p} \exp\left\{ -\frac{(\theta_1 - \theta_0)}{\sigma^2/n} \left(\frac{1}{2}(\theta_1 + \theta_0) - \bar{x} \right) \right\} + 1 \right)^{-1} \in (0,1).$$

Plugging this expression into the expected profit of exploiting the oil field and solving for $\bar{x}$ shows that you should exploit it if, and only if,

$$\bar{x} > \frac{1}{2}(\theta_1 + \theta_0) + \frac{\sigma^2/n}{\theta_1 - \theta_0} \log\left(\frac{1-p}{p} \frac{I - \theta_0}{\theta_1 - I} \right).$$

I hope it is clear by now that the Bayesian approach is very versatile. What I like about it is that it allows you to separate the *evidential* side of statistics from the *consequences* of decisions. If the purpose of your analysis is simply to report the evidence that your data contains, you can present, for example, an HDR (possibly for different priors). If a decision maker wants to use that information to inform a decision-making process, then you can use the decision theoretic framework to correctly use evidence to reach a decision. As we have seen: different purposes lead to different decision rules and thus to potentially different decisions.

10.9 Linear regression

To finish this book, let's briefly look at the normal linear regression model from a Bayesian perspective. Recall that, in that model, we assume that two random variables X and Y are linked linearly:

$$Y|\{X = x\} \sim N(\beta_0 + \beta_1 x, \sigma^2),$$

for unknown parameters β_0 and β_1. Typically, σ is also unknown and enters the analysis as a nuisance parameter.

In the Bayesian approach we view all three parameters as realizations of three independent random variables. Typically, one chooses vague priors for β_0 and β_1, while the prior for σ^2 is assumed to be inversely related to σ^2, i.e.,

$$p(\beta_0, \beta_1, \sigma^2) \propto \frac{1}{\sigma^2}.$$

Denote the sample $(y_1, x_1), \ldots, (y_n, x_n)$ by $(\mathbf{y}, \boldsymbol{x})$. Applying Bayes' rule, it can now easily be seen that the posterior distribution has density

$$p(\beta_0, \beta_1, \sigma^2 | (\mathbf{y}, \boldsymbol{x}) \propto \frac{1}{\sigma^2} \frac{1}{\sqrt{2\pi\sigma^2}} \exp\left\{-\frac{1}{2\sigma^2} \sum_{i=1}^{n} (y_i - \beta_0 - \beta_1 x_i)^2\right\}.$$

From this joint distribution, the (marginal) distributions of $\tilde{\beta}_0|(\mathbf{y}, \boldsymbol{x})$ and $\tilde{\beta}_1|(\mathbf{y}, \boldsymbol{x})$ can be found by integrating out (β_1, σ^2) and (β_0, σ^2), respectively. It can be shown (although this is not straightforward and goes too far for us here) that the posterior distributions of $\tilde{\beta}_0$ and $\tilde{\beta}_1$ satisfy

$$\frac{\tilde{\beta}_0 - \beta_0^{OLS}}{\mathsf{SE}(\beta_0^{OLS})} \sim t_{n-2} \quad \text{and} \quad \frac{\tilde{\beta}_1 - \beta_1^{OLS}}{\mathsf{SE}(\beta_1^{OLS})} \sim t_{n-2},$$

where β_0^{OLS} and β_1^{OLS} are the *realizations* of the OLS estimators $\hat{\beta}_0^{OLS}$ and $\hat{\beta}_1^{OLS}$, respectively, and the standard errors are computed as in Section 9.4.

These results are very similar to the frequentist results on linear regression, but note that the frequentist regards the parameters as constant and the estimators as random, whereas the Bayesian views the parameter as random and the estimate as given. Also, whereas the frequentist starts by deriving an estimator to base her inferences on, the Bayesian proceeds straight to the posterior distribution. When forced, a Bayesian could always come up with a (point) estimate of the unknown parameter by taking the expectation of the posterior distribution. In the case of linear regression with the particular prior used here, those estimates would coincide with the OLS estimates that the frequentist would compute.

10.10   Chapter summary

In this chapter we briefly looked at how we can draw inferences according to the Bayesian school. We saw that Bayesians use the data to update a prior distribution over the unknown parameter into a posterior distribution by appealing to Bayes' rule. We developed some ideas that could be seen as Bayesian equivalents of confidence intervals and hypothesis tests. We also briefly discussed the likelihood approach to statistical inference and embedded the Bayesian approach within a decision theoretic framework to link inferences to decision making. Finally, we discussed a Bayesian version of the normal linear regression model.

10.11   Exercises and problems

Exercises

Exercise 10.1. Suppose that you have a statistical model $(X_i)_{i=1}^n \overset{iid}{\sim} \mathsf{Bern}(\theta)$ and a prior $\tilde{\theta} \sim \mathsf{Beta}(a,b)$ with a beta distribution.[7]

(a) Show that the beta distribution is a conjugate prior for the Bernoulli likelihood.

(b) Find the mean and variance of the posterior distribution of $\tilde{\theta}$.

(c) Use the quadratic loss function to find the Bayes estimator for θ.

For $\theta \in (0,1)$, define the **odds** of a success by

$$\lambda := \frac{\theta}{1-\theta}.$$

It turns out that, if $\tilde{\theta} \sim \mathsf{Beta}(a,b)$, it holds that $\frac{b}{a}\tilde{\lambda} \sim F_{2a,2b}$.

(d) Find a $1-\alpha$ highest-density region for $\tilde{\lambda}$.

It is not necessarily obvious how to choose the prior parameters a and b. It has been argued (see, for example, Lee, 2012, Section 3.1) that you should choose a and b such that $\mathsf{E}(\tilde{p})$ equals your prior expectation and $a+b$ equals the number of observations you think this prior information is worth.

---

[7]The beta distribution is very flexible and can describe almost any distribution over the unit interval that you could wish, apart from multi-modal ones.

Exercise 10.2. Suppose that $(X_i)_{i=1}^n \overset{iid}{\sim} \mathsf{N}(\mu, \sigma^2)$, with σ known, and consider the hypotheses

$$H_1 : \mu = \mu_1 \quad \text{and} \quad H_2 : \mu = \mu_1 + \delta,$$

with known values for μ_1 and $\delta > 0$.

(a) Use (10.9) to derive (10.10).

(b) Derive the missing steps in the calculation of W_1 in (10.11).

(c) Show that $W_2 = W_1$.

(d) Derive the missing steps in the calculations of M_2 and M_1 in (10.12) and (10.13), respectively.

Exercise 10.3. Consider the problem discussed in Section 10.8. There we assumed that the parameter θ could only take two values. In this exercise we will generalize to a model where θ can take any positive value. So, consider the same model, but with the following modifications:

- the parameter space is $\Theta = (0, \infty)$;

- the utility function is $U(a_1, \theta) = \theta - I$ and $U(a_0, \theta) = 0$, for all $\theta \in \Theta$;

- the prior is $\tilde{\theta} \sim \mathsf{G}^{-1}(\alpha, \beta)$, the **inverse gamma** distribution, with density

$$p(\theta) = \frac{\beta^\alpha}{\Gamma(\alpha)} \theta^{-\alpha-1} e^{-\beta/\theta};$$

- the sample is drawn from the exponential distribution: $x|\theta \sim \mathsf{Exp}(\theta)$.

Using the same procedure as in Section 10.8, find the decision rule, based on an appropriately chosen sufficient statistic, that maximizes utility. [Hint: The inverse gamma distribution is a conjugate prior. You will also have to compute (or find) $\mathsf{E}(X)$.]

Problems

The model set-up in following problem has been taken from Berger (1985, Exercise 4.62).

Problem 10.1. You are the manager of a firm that regularly receives shipments from a supplier. A fraction θ is defective and you have to decide for each shipment whether you want to send it back to the supplier or not. The parameter space for this situation is $\Theta = (0, 1)$ and the action space is $\mathscr{A} = \{a_1, a_2\}$, where a_1 (a_2) denotes accepting (rejecting) the shipment. You have prior beliefs that $\theta \sim \mathsf{Beta}(1, 9)$. As the loss function, you take

$$L(a_1, \theta) = 100\theta \quad \text{and} \quad L(a_2, \theta) = 1.$$

You take a sample without replacement from each shipment of n items. Assuming that n is small relative to the shipment size, an appropriate statistical model is $(X_i)_{i=1}^n \overset{iid}{\sim} \mathsf{Bern}(\theta)$.

(a) Derive the posterior distribution $\tilde{\theta}|\boldsymbol{x}$.

(b) Find the Bayes rule.

Problem 10.2. The marketing officer of your cosmetics company has conducted a survey among buyers of your skin cleaning product and asked participants whether they would buy a new product your company might develop. Out of 49 respondents, 29 indicated they might buy the new product.

(a) Give some reasons why your sample may not constitute a random sample.

For the remainder of this exercise, assume that $(X_i)_{i=1}^{49} \overset{iid}{\sim} \mathsf{Bern}(\theta)$.

(b) Report to management on the research, assuming that you are a frequentist statistician.

(c) Report to management on the research, assuming that you are a Bayesian statistician who thinks a priori that the expectation of $\tilde{\theta}$ is 50%, that the prior information is "worth" six observations, and that $\tilde{\theta} \sim \mathsf{Beta}(a,b)$. [Hint: see Exercise 10.1.]

The reason that this research was conducted is because management is considering investing \$0.4 mln in a marketing campaign, which is expected to increase your sales. It is assumed that 70% of those who say they might buy the product will actually buy it. The total market consists of 1 mln potential customers and the profit margin on each unit sold is \$20. The prior information given in (c) has been obtained from sources close to management.

(d) Formulate the situation described above as a statistical decision problem.

(e) Before the results of the sample are in, advise management on whether the marketing campaign should be run or not.

(f) Before the results of the sample are in, but after you know that you will observe a random sample $(X_i)_{i=1}^{49} \overset{iid}{\sim} \mathsf{Bern}(\theta)$, derive the Bayes rule in terms of $\sum_{i=1}^n x_i$.

(g) Combining prior information and sample results, advise management on whether the marketing campaign should be run or not.

Appendices

Appendix A

Commonly used discrete distributions

Discrete random variables take values in a countable set. The **Bernoulli** distribution is used to model experiments where there are only two possible outcomes: "success" (outcome 1) or "failure" (outcome 0). The probability of success is p, which is the parameter of this distribution. When counting the number of successes in n independent Bernoulli experiments, this number follows a **binomial** distribution. This is equivalent to an experiment where you draw n items from a population *with replacement* and count the number of draws that exhibit a certain characteristic. If you draw n times *without replacement* from a population with size N in which a fraction p of members have the characteristic, you should model the number of successes using the **hypergeometric** distribution. For N large, the binomial distribution $\text{Bin}(n, p)$ gives a good approximation, which is often much easier to use.

The **Poisson** distribution is often used to model the number of occurrences of a phenomenon of which the probability of occurrence in a unit of time is small and the probability of more than one occurrence in a unit of time is negligible. Occurrences in disjunct units of time are assumed to be independent. The units of time are not necessarily small. For example, the number of traffic accidents per month or the number of stock market crashes per decade can be modeled using a Poisson distribution.

The **geometric** distribution can be used if we are interested in an experiment where we count the number of failures before the first success is observed. If we repeat this experiment n times, we get the **negative binomial** distribution.

TABLE A.1: Commonly used discrete distributions.

Distribution		domain	parameter	$E(X)$	$\mathrm{Var}(X)$	$f_X(k)$	$\sum_{i=1}^n X_i$
Bernoulli	$\mathrm{Bern}(p)$	$\{0,1\}$	$0<p<1$	p	$p(1-p)$	$p^k(1-p)^{1-k}$	$\mathrm{Bin}(n,p)$
Binomial	$\mathrm{Bin}(n,p)$	$\{0,1,\ldots,n\}$	$\left(\begin{smallmatrix}n\in\mathbb{N}\\0<p<1\end{smallmatrix}\right)$	np	$np(1-p)$	$\binom{n}{k}p^k(1-p)^{n-k}$	
Hypergeometric	$\mathrm{H}(n,N,p)$	$\{0,1,\ldots,n\}$	$\left(\begin{smallmatrix}n,N\in\mathbb{N},p\in(0,1),\\n\leq N\end{smallmatrix}\right)$	np	$\frac{nNp(N-n)N(1-p)}{N^2(N-1)}$	$\frac{\binom{Np}{k}\binom{N(1-p)}{n-k}}{\binom{N}{n}}$	
Poisson	$\mathrm{Poiss}(\lambda)$	$\{0,1,\ldots\}$	$\lambda>0$	λ	λ	$\frac{\lambda^k}{k!}e^{-\lambda}$	$\mathrm{Poiss}(n\lambda)$
Geometric	$\mathrm{Geo}(p)$	$\{0,1,\ldots\}$	$0<p<1$	$\frac{1-p}{p}$	$\frac{1-p}{p^2}$	$p(1-p)^k$	
Negative Binomial	$\mathrm{NB}(n,p)$	$\{0,1,\ldots\}$	$\left(\begin{smallmatrix}n\in\mathbb{N}\\0<p<1\end{smallmatrix}\right)$	$n\frac{1-p}{p}$	$n\frac{1-p}{p^2}$	$\binom{n+k-1}{n-1}p^n(1-p)^k$	$\mathrm{NB}(n,p)$

Appendix B

Commonly used continuous distributions

The **normal** distribution is the most often used distribution in statistics. It can be used to model experiments where the observations are clustered around the mean with exponentially decaying tails. The normal distribution is also important because the central limit theorem (CLT) says that, for large samples, the standardized sample mean approximately follows a standard normal distribution, no matter what the distribution of the underlying random variable is (as long as it has a finite mean and variance).

If the outcomes of an experiment are, by definition, between two points and each point in between is equally likely to be observed, then the **uniform** distribution can be used.

In describing waiting times and life cycles, we often use the **exponential** distribution. One of the important properties of this distribution is that it is *memoryless*, i.e., if $X \sim \mathsf{Exp}(\lambda)$, then $\mathsf{P}(X > t + u | X > u) = \mathsf{P}(X > t)$. This implies that, if X denotes the time of failure of some gadget, then a used gadget is as good as a new one. This may not be realistic in a model for life cycles, but it could be appropriate for waiting times. Also, if a random variable follows a $\mathsf{Poiss}(\lambda)$ distribution, then the inter-arrival times of occurrences follow an exponential distribution. If the total life span of a number of items is measured, each of which follows an exponential distribution and is independent of the other, then we get the **gamma** distribution.

The **Pareto** distribution is often used to model situations where a lot of observations are clustered around certain values, but with significant and large deviations. The most famous example is the income distribution. There are a lot of people whose incomes are around the average/median income, but a few very wealthy individuals. The **beta** distribution can be seen as a generalization of the $\mathsf{U}(0, 1)$ distribution. It is often used in Bayesian statistics.

The **gamma function**, $\Gamma(\cdot)$, which is encountered in several distributions described in Table B.1, has the following properties:

$$\Gamma(x) := \int_0^\infty t^{x-1} e^{-t} dt \quad (x > 0);$$
$$\Gamma(x + 1) = x\Gamma(x) \quad (x > 0)$$
$$\Gamma(n) = (n - 1)! \quad (n \in \mathbb{N}).$$

TABLE B.1: Commonly used continuous distributions.

Distribution		domain	parameter	$E(X)$	$Var(X)$	$f_X(x)$	$\sum_{i=1}^n X_i$
Uniform	$U(a,b)$	(a,b)	$a<b$	$\frac{a+b}{2}$	$\frac{(b-a)^2}{12}$	$\frac{1}{b-a}$	
Normal	$N(\mu,\sigma^2)$	$\mathbb{R}$	$(\mu\in\mathbb{R}, \sigma>0)$	μ	σ^2	$\frac{1}{\sigma\sqrt{2\pi}}e^{-\frac{1}{2}\left(\frac{x-\mu}{\sigma}\right)^2}$	$N(n\mu, n\sigma^2)$
Exponential	$Exp(\lambda)$	$(0,\infty)$	$\lambda>0$	λ	λ^2	$\frac{1}{\lambda}e^{-x/\lambda}$	
Gamma	$G(n,\lambda)$	$(0,\infty)$	$n\in\mathbb{N}, \lambda>0$	$n\lambda$	$n\lambda^2$	$\frac{1}{\lambda^n\Gamma(n)}x^{n-1}e^{-x/\lambda}$	$G(n,\lambda)$
Pareto	$Par(c,\alpha)$	(c,∞)	$c,\alpha>0$	$c\frac{\alpha}{\alpha-1}$ $(\alpha>1)$	$c^2\frac{\alpha}{(\alpha-2)(\alpha-1)^2}$ $(\alpha>2)$	$\frac{\alpha}{c}\left(\frac{c}{x}\right)^{\alpha+1}$	
Beta	$Beta(\alpha,\beta)$	$(0,1)$	$\alpha,\beta>0$	$\frac{\alpha}{\alpha+\beta}$	$\frac{\alpha\beta}{(\alpha+\beta)^2(\alpha+\beta+1)}$	$\frac{\Gamma(\alpha+\beta)}{\Gamma(\alpha)\Gamma(\beta)}x^{\alpha-1}(1-x)^{\beta-1}$	
Cauchy	$Cau(\alpha,\beta)$	$\mathbb{R}$	$(\alpha\in\mathbb{R}, \beta>0)$	undefined	undefined	$\frac{1}{\pi}\frac{\beta}{\beta^2+(x-\alpha)^2}$	

Appendix C

Special distributions

A number of distributions are encountered often in statistics. These distributions and some of their properties are very briefly described in this appendix.

C.1 χ^2 distribution

A random variable X follows the **chi-square**, or χ^2, **distribution** with ν degrees of freedom, denoted by $X \sim \chi^2_\nu$, if $X \sim G(\nu/2, 1/2)$. The following results are useful.

- Let $X \sim \text{Exp}(\lambda)$. Then $X \sim G(1, \lambda)$.

- Let $X \sim G(n, \lambda)$. Then $2X/\lambda \sim \chi^2_{2n}$.

- Let $(X_i)_{i=1}^n \overset{iid}{\sim} \chi^2_{\nu_i}$, $i = 1, \ldots, n$. Then $\sum_{i=1}^n X_i \sim \chi^2_{\sum_{i=1}^n \nu_i}$.

- Let $Z \sim N(0, 1)$. Then $Z^2 \sim \chi^2_1$.

- Let $(X_i)_{i=1}^n \overset{iid}{\sim} N(\mu, \sigma^2)$. Then $n\frac{(\bar{X}-\mu)^2}{\sigma^2} \sim \chi^2_1$.

- Let $(X_i)_{i=1}^n \overset{iid}{\sim} N(\mu, \sigma^2)$ and let $\hat{\sigma}^2_X$ and S^2_X be the sample variance and unbiased sample variance, respectively. Then $n\frac{\hat{\sigma}^2_X}{\sigma^2} = (n-1)\frac{S^2_X}{\sigma^2} \sim \chi^2_{n-1}$.

C.2 Student's t-distribution

Let Z and V be two independent random variables such that $Z \sim N(0, 1)$ and $V \sim \chi^2_\nu$. Then the random variable $T = \frac{Z}{\sqrt{V/\nu}}$ has a **t-distribution** (or Student t-distribution) with ν degrees of freedom, denoted by $T \sim t_\nu$. A useful result is the following:

- Let $(X_i)_{i=1}^n \overset{iid}{\sim} N(\mu, \sigma^2)$. Then $\frac{\bar{X}-\mu}{S_X/\sqrt{n}} \sim t_{n-1}$.

[Note where you use n and where $n - 1$; make sure you understand why.]

C.3   *F*-distribution

Let V_1 and V_2 be two independent random variables such that $V_1 \sim \chi^2_{\nu_1}$ and $V_2 \sim \chi^2_{\nu_2}$. Then the random variable $Y = \frac{V_1/\nu_1}{V_2/\nu_2}$ follows the **F-distribution** with ν_1 and ν_2 degrees of freedom, denoted by $Y \sim F_{\nu_1, \nu_2}$.

Appendix D

The Greek alphabet

Letter	Lower case	Upper case	Letter	Lower case	Upper case
Alpha	α		Nu	ν	
Beta	β		Xi	ξ	
Gamma	γ	Γ	Omicron	o	
Delta	δ	Δ	Pi	π	Π
Epsilon	ε		Rho	ρ	
Zeta	ζ		Sigma	σ	Σ
Eta	η		Tau	τ	
Theta	θ	Θ	Upsilon	υ	Υ
Iota	ι		Phi	φ	Φ
Kappa	κ		Chi	χ	
Lambda	λ	Λ	Psi	ψ	Ψ
Mu	μ		Omega	ω	Ω

Appendix E

Mathematical notation

Basic notation

$\cdot = \cdot$	the expression/number/$\cdots$ on the left is exactly equal to the expression/number/$\cdots$ on the right.
$\cdot := \cdot$	the expression on the left is defined to be equal to the expression on the right.
$\cdot \propto \cdot$	the expression on the left is proportional to the expression on the right.
$a \le b$	the number a is no larger than the number b.
$a < b$	the number a is (strictly) smaller than the number b.
$a \ge b$	the number a is no smaller than the number b.
$a > b$	the number a is (strictly) larger than the number b.
$\mathscr{A} \Rightarrow \mathscr{B}$	if statement $\mathscr{A}$ is true, then so is statement $\mathscr{B}$ ($\mathscr{A}$ implies $\mathscr{B}$).
$\mathscr{A} \iff \mathscr{B}$	statement $\mathscr{A}$ implies statement $\mathscr{B}$ and statement $\mathscr{B}$ implies statement $\mathscr{B}$, i.e., $\mathscr{A}$ is true if, and only if, B is true, i.e., $\mathscr{A}$ and $\mathscr{B}$ are equivalent.

Sets

$\emptyset$	the empty set.
$\mathbb{N}$	the set of all natural numbers.
$\mathbb{Z}$	the set of all (positive and negative) integers.
$\mathbb{Q}$	the set of all rational numbers.
$\mathbb{R}$	the set of all real numbers.
$A = \{a_1, a_2, \dots\}$	A is a set that contains all elements a_1, a_2, etc.
A^c	all elements that do not belong to the set A (the **complement** of A).
$a \in A$	the object a is an element of the set A,
$A \cup B$	the set of all elements that are in A, or in B, or in both.
$A \cap B$	the set of all elements that are in A and in B.
$A \backslash B$	the set of all elements that are in A, but not in B.
$A \cap B = \emptyset$	the sets A and B are **mutually exclusive**.
$A \cup B = \Omega$	the sets A and B are **collectively exhaustive** (for Ω).

$\{a \in A | \mathscr{A}\}$ the set of all elements of set A for which statement $\mathscr{A}$ is true.

functions

$f : A \to B$ the function f assigns to each element of the set A an element in the set B.

$x \mapsto f(x)$ the function f assigns to each x the number $f(x)$.

$\max_{x \in A} f(x)$ find the highest number that the function f can take over all elements x in the set A.

$1_A(x)$ **indicator function**: takes the value 1 if $x \in A$ and 0 otherwise.

$\arg\max_{x \in A} f(x)$ find the value(s) of x in the set A that attain $\max_{x \in A} f(x)$.

$f'(x)$ derivative of function f at x.

$\frac{\partial f}{\partial x}$ derivative of function f with respect to the variable x.

Appendix F

Summation and product notation

In statistics you often have to sum over a number of elements. For example, you might want to compute the sum of the first 10 natural numbers. You could write this, obviously, as

$$1 + 2 + 3 + 4 + 5 + 6 + 7 + 8 + 9 + 10 = 55.$$

If the number of elements to be added is large, writing every element down becomes rather cumbersome. Therefore, we often write this sum as

$$1 + 2 + \cdots + 10 = 55.$$

An even more elegant way to write it uses a standard mathematical notation, namely, the Greek capital letter "Sigma" (S for "sum"):

$$\sum_{i=1}^{10} i = 55.$$

You should read this as: "The sum of all i, where i runs from 1 to 10." The symbol i is called the *summation index*.

More generally, suppose that I have a sequence of numbers $x_1, x_2, \ldots, x_n$. For example, suppose I have a sample containing n observations on exam results of first-year university students. In that case, x_i denotes the exam result for student i in my sample. Often we want to compute the average value, which we denote by $\bar{x}$. In order to do so, we need the sum of all observations and divide them by the number of observations.

A convenient way of writing the sum is by using the $\sum$ notation:

$$\sum_{i=1}^{n} x_i := x_1 + x_2 + \cdots + x_n.$$

This should be read as: "the sum of all values x_i, where i runs from 1 to n." So, the average score of the sample of first-year university students' exam results can now be written as

$$\bar{x} = \frac{1}{n} \sum_{i=1}^{n} x_i = \frac{x_1 + \cdots + x_n}{n}.$$

By the way, you do not have to use i as the summation index. By convention, we use i as the first choice, but there is nobody stopping you from using j, or k, or l, or a, or..... In fact,

$$\sum_{i=1}^{n} x_i = \sum_{j=1}^{n} x_j = \sum_{a=1}^{n} x_a.$$

The summation index is just there to keep track of what elements we are taking the summation of. The only thing that you have to keep in mind is that you have to be *consistent*. For, example, you cannot say "I am going to sum over j" and then use k as a subscript:

$$\sum_{j=1}^{n} x_k \neq \sum_{j=1}^{n} x_j.$$

Summation notation is particularly useful if you want to sum over elements of a *set*. For example, suppose that $S = \{(H, H), (T, T), (H, T), (T, H)\}$. We will encounter this set in probability theory. It represents the list of all possible outcomes of two coin flips: (Heads, Heads), (Tails, Tails), (Heads, Tails), and (Tails, Heads). Suppose that you are engaged in a bet involving two coin flips: you win \$1 each time Heads comes up, and you lose \$1 if Tails comes up. Denote the amount won by X. Then X is a function that maps the list of all possible outcomes, S, to the set $\{-2, 0, 2\}$, since

$$X(s) = \begin{cases} -2 & \text{if } s = (T, T), \\ 0 & \text{if } s = (H, T), \text{ or } s = (T, H), \\ 2 & \text{if } s = (H, H). \end{cases}$$

The sum of all possible payoffs can now be written as

$$\sum_{s \in S} X(s) = X(T, T) + X(H, T) + X(T, H) = X(H, H) = 0,$$

which should be read as "the sum over all values $X(s)$, where s is some element of the set S." Note that here the summation index is an element s of the set S, i.e., a pair of outcomes (s_1, s_2).

In addition, you can write down the sum over all elements of a *subset* of a set. Suppose that I want to write down the sum of the payoffs of all pairs of coin flips which show Heads for the first flip. That is, I am interested in the subset $A = \{s \in S : s_1 = H\}$. This should be read as: "the set containing all elements of S for which the first element is H." In other words, $A = \{(H, H), (H, T)\}$. So,

$$\sum_{s \in A} X(s) = \sum_{\{s \in S : s_1 = H\}} X(s) = X(H, H) + X(H, T) = 2.$$

Rules for summation notation are straightforward extensions of well-known properties of summation. For example,

$$\sum_{i=1}^{n} a x_i = a x_1 + a x_2 + \cdots + a x_n$$

$$= a(x_1 + x_2 + \cdots + x_n)$$

$$= a \sum_{i=1}^{n} x_i.$$

That is, you can take a constant "out of the summation." This is nothing more than taking a constant out of brackets.

For products we do something similar, but we use the letter Π (Pi) to denote products. So,

$$\prod_{i=1}^{n} a_i := a_1 a_2 \cdots a_n.$$

Standard rules of products apply to this notation as well. For example, if $\lambda > 0$ is a constant, then

$$\prod_{i=1}^{n} \frac{1}{\lambda} = \frac{1}{\lambda^n}.$$

Appendix G

Exponentials and (natural) logarithms

The exponential function is given by the mapping $x \mapsto e^x$, which is also written as $\exp(x)$. Here e is Euler's number, $e \approx 2.718281\ldots$. The exponential function can be written as an infinite sum:

$$e^x = \sum_{k=0}^{\infty} \frac{x^k}{k!}.$$

It holds that

$$e^x > 0 \text{ (all } x), \quad \lim_{x \downarrow -\infty} e^x = 0, \quad e^0 = 1, \quad e^x \to \infty \text{ (as } x \uparrow \infty), \quad \frac{d}{x} e^x = e^x.$$

The last property implies that e^x is a strictly increasing convex function.
For all a and b it holds that

$$e^{a+b} = e^a e^b, \quad e^{a-b} = \frac{e^a}{e^b}, \quad (e^a)^b = e^{ab}.$$

The inverse of $\exp(x)$ is the natural logarithm, denoted by $\log(x)$, or $\ln(x)$. So, for all x it holds that

$$\log(e^x) = x.$$

Note that $\log(x)$ is only defined on $(0, \infty)$ (because $e^x > 0$ for all x). It holds that

$$x \to -\infty \text{ (as } x \downarrow 0), \quad \log(1) = 0, \quad \log(x) \to \infty \text{ (as } x \uparrow \infty), \quad \frac{d}{x} \log(x) = \frac{1}{x}.$$

The last property implies that $\log(x)$ is a strictly increasing concave function.
For all a and b it holds that

$$\log(ab) = \log(a) + \log(b), \quad \log\left(\frac{a}{b}\right) = \log(a) - \log(b), \quad \log\left(a^b\right) = b\log(a).$$

Appendix H

Subjective probability

So far we have interpreted probabilities as frequencies. If we say that the probability of an event is 0.8, we mean that the fraction of subjects in the population having the characteristic described in the event is 80%. This works well in some applications, such as trials for new drugs. There we would model the efficacy of a treatment in terms of quality adjusted life years (QALY) as, for example, $X \sim \text{Exp}(\lambda)$. If we take a very large sample out of the population of all possible patients, we would find, on average, a QALY gain of λ. This allows us to say that, if we choose one patient at random, our expectation is a QALY gain of λ.

The frequency interpretation is based on the law of large numbers and requires, to be accurate, a very large population.[1] So, the frequentist probability interpretation assumes a hypothetical infinitely large population in order for the frequency interpretation to make sense. In many cases that is unrealistic. When we talk about sporting events, for example, there is no infinite population. If you talk about the probability that athlete X will win the 100 m sprint at the next Olympics, there is a population of 1: the next Olympics will be played once. A frequency interpretation of the statement "the probability that athlete X will win the 100 m sprint in the next Olympics is 0.6" would be that "in 60% of all possible next Olympics, athlete X will win the 100 m sprint." That does not sound like a very useful statement.

In situations like the one described above, we use probability more as a **degree of belief**, rather than a frequency. Stock market gurus pontificating on TV about future performance of stocks tend to have a degree of belief interpretation. The question that naturally arises is: can we model "degree of belief" as a (mathematical) probability? Many scholars have thought about this issue, but the name I would like to mention here is Bruno de Finetti (1906–1985). If nothing else, writing a two-volume book called *Theory of Probability* and to use PROBABILITY DOES NOT EXIST as pretty much your first sentence (and written in capitals) takes some chutzpah!

What de Finetti (1937) tried to do was to formalize the notion of "degree of belief" and to find properties under which these admit a representation as a (mathematical) probability as defined in Chapter 2. The way de Finetti did this was by taking as his basic starting point *monetary bets on states of*

---

[1] Actually, it is a bit odd that you use a result from probability to define an interpretation of probability.

nature. The easiest way to understand these ideas is to think of a horse race. Suppose there are n horses in a race and only the winner pays out. Then a monetary bet is a vector $x = (x_1, \ldots, x_n)$, where $x_i > 0$ ($x_i < 0$) denotes the amount of money you'll win (have to pay) if horse i wins. Let $X \subset \mathbb{R}^n$ denote the set of all possible bets.

The idea is that you (individually) have **preferences** over these bets. In particular, if I show you two bets, x and x', you should be able to tell me which one you (weakly) prefer. Suppose that you weakly prefer x to x'. We write this as $x \succeq x'$. The symbol $\succeq$ means "at least as good as." We now discuss several properties that your preferences could satisfy. Their formulation closely follows Gilboa (2009). The relation "strictly preferred to" is denoted by $\succ$.

(F1) $\succeq$ is a *weak order.*

By this we mean that, for all $x, x', x'' \in X$:

1. you can make a choice between any two bets, i.e., it holds that $x \succeq x'$ or $x' \succeq x$ (we say that $\succeq$ is *complete*);

2. if you weakly prefer x to x' and x' to x'', then you must weakly prefer x to x'', i.e., $x \succeq x'$ and $x' \succeq x''$ imply $x \succeq x''$ (we say that $\succeq$ is *transitive*).

These two requirements are very natural. If the first one is violated, we can't analyse your behaviour because it is undefined for some bets. If the second property is violated, you're being inconsistent and then, again, we can't analyse anything. Transitivity is violated, for example, if you prefer apples to pears, pears to peaches, and peaches to apples. If I had to predict what you would choose if you could take one item of fruit from a selection of apples, pears, and peaches in front of you I'd have some trouble (and so would you I presume).

(F2) $\succeq$ is *continuous.*

This is a technical condition which says that, for every $x \in X$, the sets $\{y \in X | x \succ y\}$ and $\{y \in X | y \succ x\}$ are open. Condition (F2) basically means that, if you change a bet a little bit, the preference orderings do not abruptly change.

(F3) $\succeq$ is *additive.*

Your preferences are additive if, for all $x, y, z \in X$, it holds that $x \succeq y$ if, and only if, $x + z \succeq y + z$. In other words, if you prefer bet x to bet y, then the ordering is preserved if you add *the same* bet z to both x and y. Condition (F3) implies that your goal is to maximize the expected monetary value of your bets. In a pure betting situation this may be a reasonable assumption, but in many other cases it is not.[2]

---

[2]There has been a lot of research into ways to relax (F3); see, for example, Gilboa (2009).

(F4) $\succeq$ is *monotonic*.

This condition states that, if I offer you two bets x and y, such that the payoff in x is at least as high as the corresponding payoff in y, then you must prefer x to y. In mathematical notation: for every $x, y \in X$, $x_i \geq y_i$, all i, implies that $x \succeq y$. Your preferences are monotonic if you think that "more is better," which, in terms of monetary bets, does not seem a strong assumption to make.

(F5) $\succeq$ is *non-trivial*.

By this we mean that there must be at least one bet x that is *strictly* preferred to some other bet y, i.e., $x \succ y$. If this were not the case, you would be indifferent between all bets and choosing the most preferred one is impossible.

It now turns out that (F1)–(F5) are necessary and sufficient for the existence of a probability under which you can represent your preferences by the expected value of monetary bets.

Theorem H.1 (de Finetti). $\succeq$ *satisfies (F1)–(F5) if, and only if, there exists a probability vector $p \geq 0$, $\sum_{i=1}^{n} p_i = 1$, such that, for every $x, x' \in X$, it holds that*

$$x \succeq x' \iff \sum_{i=1}^{n} p_i x_i \geq \sum_{i=1}^{n} p_i x_i'.$$

This theorem says that all I need to know about your preferences is the vector p. I can then measure your valuation of any bet x by computing the expected value of x using the probabilities p. Take, for example, the situation with $n = 3$ and consider the bets $x = (1, 0, 0)$ and $x' = (0, 1, 0)$. Suppose that your preferences satisfy (F1)–(F5) and that you tell me that you prefer bet x to x'. That means that I can find a vector of probabilities p with the property that

$$\sum_{i=1}^{n} p_i x_i \geq \sum_{i=1}^{n} p_i x_i'$$

$$p_1 \geq p_2.$$

So, from your statement about which of the bets x and x' you prefer, I can *deduce* that you think it is more likely that horse 1 wins than that horse 2 wins.

This then gives a way for me to try and elicit the vector p. I will keep asking you to choose between pairs of bets. Each answer you give puts an additional constraint on the vector p. At some stage I will know the exact probabilities. In fact, I will be particularly interested in **indifferences**: those bets x and x' for which it holds that $x \succeq x'$ *and* $x' \succeq x$. We denote this by $x \sim x'$. From Theorem H.1 it immediately follows that the vector p is such that, for those bets,

$$\sum_{i=1}^{n} p_i x_i = \sum_{i=1}^{n} p_i x_i'.$$

For example, if you tell me that you are indifferent between $(0.5, -1, 0)$ and $(0, 0, 0)$, then I can immediately conclude that $p_2 = 0.5p_1$. If you also tell me that you are indifferent between $(0, 1, -1)$ and $(0, 0, 0)$, then I find that $p_3 = p_2$. Combining this information with the fact that the vector p should be a probability, I can now see that it must be the case that $p = (0.5, 0.25, 0.25)$.

Note that the probabilities provided by Theorem H.1 are completely subjective, or personal. My preferences may be different. For example, for me it might hold that

$$(1, -1, 0) \sim (0, 0, 0) \quad \text{and} \quad (0, 1, -1) \sim (0, 0, 0),$$

so that for me $p = (1/3, 1/3, 1/3)$.

References

Aitken, C., P. Roberts, and G. Jackson (2010). *Fundamentals of Probability and Statistical Evidence in Criminal Proceedings*. Royal Statistical Society.

Albert, J. (2009). *Bayesian Computation with R* (Second ed.). Springer, Dordrecht.

Berger, J. (1985). *Statistical Decision Theory and Bayesian Analysis* (Second ed.). Springer, Berlin.

Bowley, A. (1937). *Elements of Statistics*. King & Son, Westminster.

Brazzale, A., A. Davison, and N. Reid (2007). *Applied Asymptotics*. Cambridge University Press, Cambridge, UK.

de Finetti, B. (1937). La prévision: ses lois logique, ses sources subjectives. *Annales de l'Institut Henri Poincaré 7*, 1–68.

Edwards, A. (1992). *Likelihood* (Second ed.). The Johns Hopkins University Press, Baltimore.

Evans, K., P. Tyrer, J. Catalan, U. Schimidt, K. Davidson, J. Dent, P. Tata, S. Thornton, J. Barber, and S. Thompson (1999). Manual-assisted cognitive-behaviour therapy (mact): A randomized controlled trial of a brief intervention with bibliotherapy in the treatment of recurrent deliberate self-harm. *Psychological Medicine 29*, 19–25.

Fisher, R. (1973). *Statistical Methods and Scientific Inference* (Third ed.). Hafner Press, New York.

Gardner, D. (2008). *Risk: The Science and Politics of Fear*. Virgin Books, London.

Gilboa, I. (2009). *Theory of Decision under Uncertainty*. Cambridge University Press, New York.

Goldacre, B. (2012). *Bad Pharma*. Fourth Estate, London.

Hacking, I. (1965). *Logic of Statistical Inference*. Cambridge University Press, Cambridge, UK.

Jeffreys, H. (1961). *Theory of Probability* (Third ed.). Oxford University Press, Oxford.

Kahneman, D. (2011). *Thinking, Fast and Slow.* Allen Lane, London.

Keuzenkamp, H. (1995). *Probability, Econometrics, and Truth.* Cambridge University Press, Cambridge, UK.

Leamer, E. (1978). *Specification Searches.* John Wiley & Sons, New York.

Lee, P. (2012). *Bayesian Statistics. An Introduction* (Fourth ed.). John Wiley & Sons, Chichester.

Neave, H. (1978). *Statistics Tables for mathematicians, engineers, economists and the behavioural and managment sciences.* Routledge, London.

Pratt, J., H. Raiffa, and R. Schlaifer (1995). *Introduction to Statistical Decision Theory.* MIT Press, Cambridge, MA.

Royall, R. (1997). *Statistical Evidence. A Likelihood Paradigm.* Chapman & Hall/CRC, Boca Raton, Florida.

Sinn, H. (2009). *Kasino Kapitalismus. Wie es zur Finanzkrise kam, und was jetzt zu tun ist* (Second ed.). Econ Verlag, Berlin.

van der Vaart, A. (1998). *Asymptotic Statistics.* Cambridge University Press, Cambridge, UK.

Wheelan, C. (2013). *Naked Statistics: Stripping the Dread from the Data.* W. W. Norton & Company, New York.

Ziliak, S. and D. McCloskey (2008). *The Cult of Statistical Significance.* The University of Michigan Press, Ann Arbour, Michigan.

Index

action, 174
action space, 174
asymptotic sampling distribution, *see* sampling

Bayes
    decision, 175
    estimator, *see* estimator
    expected loss, 175
    factor, 168
    rule, 15, 163, 179
bias, *see* estimation
binomial coefficient, 17

cardinality, 13
central limit theorem, *see* theorem
Chebychev inequality, 23
complement, 193
confidence interval, 51, 99, 101
    based on ML estimator, 105
    definition of, 102
    for difference in mean, 107
    for difference in proportion, 108
    for mean, 50, 101, 104, 105
        properties of, 58
    for proportion, 105
    for regression coefficients, 152
    for variance, 109
confidence level, 51, 99
continuity correction, 35
convergence
    in distribution, 31
    in probability, 29
correlation, *see* random variable
covariance, *see* random variable
Cramér–Rao
    lower bound, 85, 93, 105

theorem, *see* theorem
data
    experimental, 159
    obervational, 159
decision rule, 174
decision theory, 174
degree of belief, 201
density
    function, 16
    joint, 23
    marginal, 23
distribution, 3
    χ^2, 68, 108, 189
    Bernoulli, 18, 64, 185
    beta, 187
    binomial, 18, 43, 185
    conditional, 24
    exponential, 65, 70, 114, 138, 187
    F, 153, 190
    function, 15
    gamma, 71, 187
    geometric, 95, 138, 185
    Hardy–Weinberg, 96
    hypergeometric, 19, 64, 185
    inverse gamma, 181
    memoryless, 34, 187
    multivariate, 23
    negative binomial, 185
    normal, 25–27, 64, 187
        bivariate, 27, 65
        standard, 25
    Pareto, 187
    Poisson, 94, 138, 185
    skewed, 5, 21
    Student's t, 68, 103, 189